Spherical Models

Magnus J. Wenninger

Cambridge University Press

Cambridge London New York Melbourne

Published by the Syndics of the Cambridge University Press
The Pitt Building, Trumpington Street, Cambridge CB2 1RP
Bentley House, 200 Euston Road, London NW1 2DB
32 East 57th Street, New York, NY 10022, USA
296 Beaconsfield Parade, Middle Park, Melbourne 3206, Australia

First published 1979

Printed in the United States of America
Typeset by David E. Seham Associates, Inc., Metuchen, N.J.
Printed and bound by The Murray Printing Company, Westford, Mass.
Frontispiece printed by The Longacre Press, Inc., New Rochelle, N.Y.

Library of Congress Cataloging in Publication Data

Wenninger, Magnus J.

Spherical models.

Includes bibliographical references.

1. Sphere – Models. 2. Geodesic dome – Models.
I. Title.
QA491.W43 516′.33 78–58806
ISBN 0 521 22279 6 hard covers
ISBN 0 521 29432 0 paperback

to the memory of
my father and my mother

and

to Professor H. S. M. Coxeter
without whose inspiration
this book could never have been written

Certe in Dei Creatoris mente consistit Deo coaeterno figurarum harum veritas.

Beyond doubt the true form of all these shapes exists eternally in the mind of God the Creator.

Johann Kepler, 1611

Contents

Foreword

In order to provide the reader with an indication of his standards of judgment, a book reviewer must carefully weigh and clearly state the pros and cons of any book. Unqualified praise is suspect, for even the most excellent of books will be graced by some features that will displease some reviewer.

It is very gratifying to find that a minor criticism in my review of Magnus Wenninger's superb *Polyhedron models* may in some small measure have provided an impetus for his present *Spherical models*. I am delighted to find my role changed from reviewer of the earlier volume to contributor of a foreword to the latter.

The projection of polyhedrons onto a spherical surface has distinct conceptual advantages. Although not flat, the surfaces of such polyhedrons are two-dimensional, allowing two degrees of freedom of travel on them. The spherical surface provides these polyhedrons with a standard frame of reference, having two degrees of freedom, on which to represent, transform, and interrelate these forms without change in the radial coordinates of their vertices, edges, and faces. Although the rational relationships between the volumes of some of these solids are obscured by spherical projection, edge lengths and distances obtain simple relations, expressible as arc lengths of great and small circles on the surface of the sphere. Furthermore, spherical projection relates polyhedrons to dome structures, since edges of these polyhedrons constitute or relate to geodesics, that is, to paths of minimal length between two points on the surface of a sphere.

The tessellation, or tiling, of a surface has fascinated artists and designers for many centuries. The tiling of a plane is easily related to that of a cylinder, because the plane may be rolled up into a cylinder. Certain constraints need to be specified along the seam, but basically the problem of tiling the cylinder is not very different from that of tiling the plane. In turn, a cylinder may be bent so that its ends join; the result is a toroid. Aside from another seam constraint, and some distortion resulting from the bending, the problem of tiling a toroid does not essentially differ from that of tiling the plane. Sphere tessellations, on the other hand, follow entirely different rules, which are essentially those determining polyhedral configurations. The present book, then, is a book concerned in a unified manner with polyhedrons, sphere tilings, and dome structures.

The beauty of Magnus Wenninger's models is beyond doubt. It would have been easy for him to display these "live" and through photographs, withholding the secrets of their manufacture, or to veil their mathematics in obscure formulas or symbolism, restricting their accessibility to a cultural elite. Instead the author has devised an optimum method of construction, using minimal, and principally planar, trigonometry. Teachers of design science can now provide their students with construction materials and this book and feel confident that successful models will emerge. And once these attractive models become prevalent, they are bound to influence both environmental art and architecture and, thus, have a very positive effect on our visual environment. We owe Magnus Wenninger a debt of gratitude for making his jewels, or three-dimensional mandalas, so accessible!

Arthur L. Loeb

April, 1978
Cambridge, Massachusetts

Preface

Since the publication of my book *Polyhedron models* (Cambridge University Press, 1971), one of my ambitions has been to clarify for the general reader the section entitled "Mathematical classification," pp. 4–10, of that book. Even so highly qualified a reader as Arthur L. Loeb of the Department of Visual and Environmental Studies at Harvard University commented in his review of the book: "This section suffers from trying at once too much and too little: in its conciseness it is very difficult to follow and not really extensive enough to provide a sufficient background to the relationships between the various polyhedra" (*Leonardo,* vol. 7, p. 73).

It is my hope that the present book will come to the aid of any interested reader and supply what is needed for a better understanding of polyhedral relationships. On p. 4 of *Polyhedron models* I said: "The ideas are easier to visualize with the aid of models," and the same is true here. Fortunately, the methods used for making the three spherical models illustrated on p. 7 of *Polyhedron models* can be generalized. Therefore the present book is simply entitled *Spherical models* and may be considered a companion volume to *Polyhedron models,* though by no means as comprehensive in scope.

The spherical models of this book are closely related to geodesic domes. Many people are vaguely aware of the relationship of polyhedral forms to geodesic domes. Polyhedron models generally are found to be very attractive mainly because of their symmetries. This remains equally true for the spherical models given here. It will, therefore, also be the aim of this book to show explicitly the relationship between polyhedrons and geodesic domes and to show how models of such domes can be made in paper.

Even though all the models illustrated in this book are truly spherical, it may be in place here to say immediately that the mathematics involved remains elementary throughout. The book will be directed mainly toward the use of students of mathematics and their teachers on the high school level. The beauty of these shapes, well known through the work of Buckminster Fuller, naturally will also attract students and teachers in art departments and in engineering projects. This book is not intended as an engineer's or an architect's manual. Yet it may, indeed, be used even by students in such advanced and technical departments at universities or on the college level as a source book for ideas. Many interesting spherical forms are presented here, not only the regular and semiregular shapes, but star-faced spherical models as well, all of which can serve as basic structures for geodesic domes. The book should equally well benefit interior decorators and designers of decorative devices, lighting fixtures, lampshades, and mobiles.

Although the material used for making the models shown in this book is paper alone (tag or index card stock), other more durable materials will readily suggest themselves. The advantage of paper is not only its cheapness but its easy handling. I have myself constructed all the models presented here and shown in the photographs. The working time for each is directly proportional to the number of small spherical triangles needed to complete a full spherical model, from an hour or two to fifty or a hundred hours depending on complexity.

Special thanks must be extended to Dr. John Skilling of the Department of Applied Mathematics and Theoretical Physics, University of Cambridge, England, for having provided the computer graphics called plates that appear in this book. The photography is the work of Stanley Toogood, and beyond doubt the clarity of these pictures adds much to show the models as they really are. Dr. Joseph Clinton of Kean College, New Jersey, advised me on certain details with respect to geodesic domes.

M. J. W.

Introduction: Basic properties of the sphere

All the spherical models presented in this book can be thought of as being derived from patterns formed by arcs of great circles on a sphere. The sphere itself is simply defined as the locus of points in three-dimensional space all of which are at a given distance from a fixed point called the center. The given distance is the radius of the sphere. The sphere can also be said to be the solid bounded by its spherical surface.

Some important properties of the sphere are the following:

1. Every section of a spherical surface made by a plane is a circle. In particular this means that each face of a polyhedron inscribed in a sphere lies on a plane that cuts the sphere in a circle. This circle is the circumcircle of that face and it is called a small circle. Its center is the center of the polyhedral face related to it and its radius is less than that of the sphere itself.

2. If a plane passes through the center of the sphere, the section is called a great circle. Its center and radius are the same as those of the sphere itself. In particular it should be noted that only in the case of some nonconvex uniform polyhedrons will the faces lie on planes through the center of the polyhedron. This center coincides with the center of the circumscribing sphere.

3. A unique great circle is determined by the center of the sphere and any two points on its surface provided these points are not the extremities of a diameter. In particular this means that each edge of a polyhedron determines an arc of a great circle, since the two end points of an edge lie on the surface of the sphere.

4. The shortest path from one point to another on a spherical surface is along the arc of a great circle. This shortest path is called a geodesic. In particular the edges of a polyhedron can be replaced by arcs of great circles to obtain a spherical polyhedron. Each plane polygon that is a face of the polyhedron is thus transformed into a spherical polygon that is a face of the spherical polyhedron.

Another way to think of this transformation is to imagine a polyhedron enclosed within its circumscribing sphere. By central (also called gnomonic) projection, the edges of the polyhedron generate a set of arcs of great circles on the surface of the sphere, decomposing the surface into a set of spherical polygons, one for each face of the polyhedron. If the planes of symmetry of the polyhedron are now introduced, each plane cuts the sphere in a great circle, because these planes all pass through the center of the polyhedron that is also the center of the sphere. The intersections of all of these great circles decompose the surface of the sphere into a network of spherical triangles. This is called spherical tessellation. It is from these spherical triangles that a spherical model can be made.

With this much abstract theory now set down (more will come later), it is time to turn to its practical application. At first you might feel that a great deal of spherical trigonometry is involved in finding the angle measures of all the spherical triangles needed for a spherical model. Fortunately this is not the case. If you know some elementary plane geometry, if you have done some constructions with ruler and compass, you will be able to follow the steps set out in this book to make not only the three regular spherical models but also the full set of all thirteen semiregular spherical ones as well. And as a bonus you receive the duals at the same time. Duals will be explained later.

Beyond the plane geometry, a knowledge of plane trigonometry, using the sine, cosine, and tangent functions, is all you will need to provide the background for angle measures to check the accuracy of the geometrical constructions. The use of a small hand-held electronic calculator, the slide rule or scientific type, is highly recommended here, because it takes the tedium out of paper and pencil calculations using mathematical tables or even the slide rule, while it gives you abundant practice in deriving results that are immediately useful, namely answers that will be used for real work – not just answers alone.

You will also find that by the use of ordinary geometry and trigonometry you can make some spherical models of star-faced polyhedrons. For example a five-pointed star called a pentagram can be drawn on each face of a dodecahedron. Eight-pointed stars called octagrams and ten-pointed stars called decagrams can be given the same treatment on faces of the respective polyhedrons that have octagons and decagons for faces.

Furthermore, some variations may be introduced for the regular or semiregular spherical models so that they take on the appearance of genuine geodesic domes. This is a particularly delightful result to come upon, because it leads to the next section of this book, how to make paper models of geodesic domes, as complex as may be desired. Here some spherical trigonometry will be introduced, but the calculations can very easily be done on a small electronic calculator. You can make a complete spherical model of a dome composed of 1,280 spherical triangles with no great trouble except that of persevering to complete it.

So begin by making a few of the simple ones first. Once you see how beautiful they are you will be moving on from one to the next, each one giving you a delight all its own. It is really surprising how much enlightenment will come following the construction of the models rather than preceding it. It is highly recommended that in each case you make your own drawings.

I. The regular spherical models

The basic symmetry groups of the five regular polyhedrons actually reduce themselves to three – the tetrahedral, the octahedral, and the icosahedral. You may in fact already be aware that the tetrahedron, the octahedron, and the icosahedron all have faces that are equilateral triangles. Projecting their edges onto the surface of their respective circumscribing spheres generates in each case a set of 4, 8, and 20 equilateral spherical triangles. Introducing the respective planes of symmetry for each yields the complete set of 24, 48, and 120 right spherical triangles. These are illustrated in Photos 1, 2–3, and 4–5.

Photo 2–3. Octahedron; cube.

Photo 1. Tetrahedron.

Photo 4–5. Icosahedron; dodecahedron.

The spherical hexahedron or cube

To learn how to make these spherical models you will find it easiest if you begin with the cube (see Fig. 1). Study carefully the shape shown in heavy lines on the interior of the cube, which is shown with its circumscribing sphere. This shape is a special type of tetrahedron called an orthoscheme or a quadrirectangular tetrahedron because each of its four faces is a right-angled triangle. The cube has forty-eight such tetrahedral cells in mirror-image pairs. In Fig. 1 you see that O_0 is a vertex of cube, O_1 is the midpoint of an edge, O_2 is the incenter of a face, and O_3 is the center of the cube that is at the same time the center of the circumscribing sphere. Triangle $O_0O_1O_2$ on the face of the cube can be projected by central or gnomonic projection onto the surface of the sphere, generating the spherical triangle $Q_0Q_1Q_2$. Notice that Q_0 and O_0 are the same point. If the cube is now assigned an edge length of $e = 2$, then half of e equals 1 (one unit of length). This is the length of O_0O_1.

It is from the spherical triangle $Q_0Q_1Q_2$

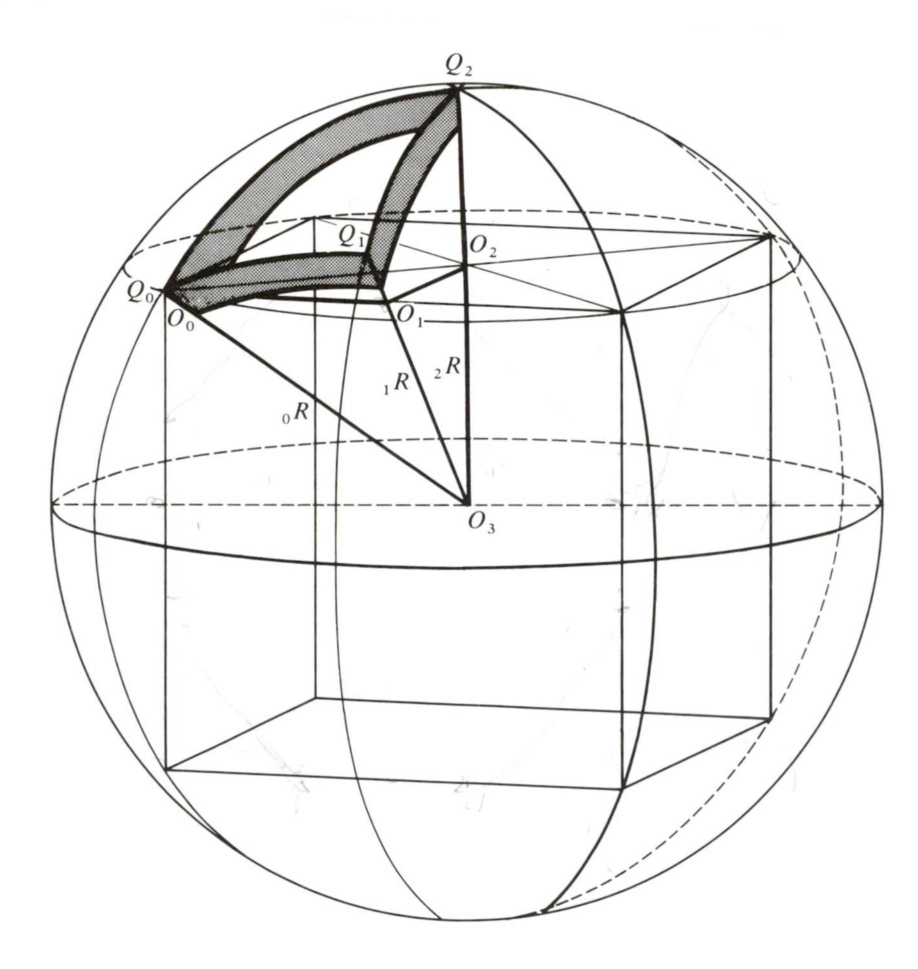

Fig. 1. Cube with circumscribing sphere.

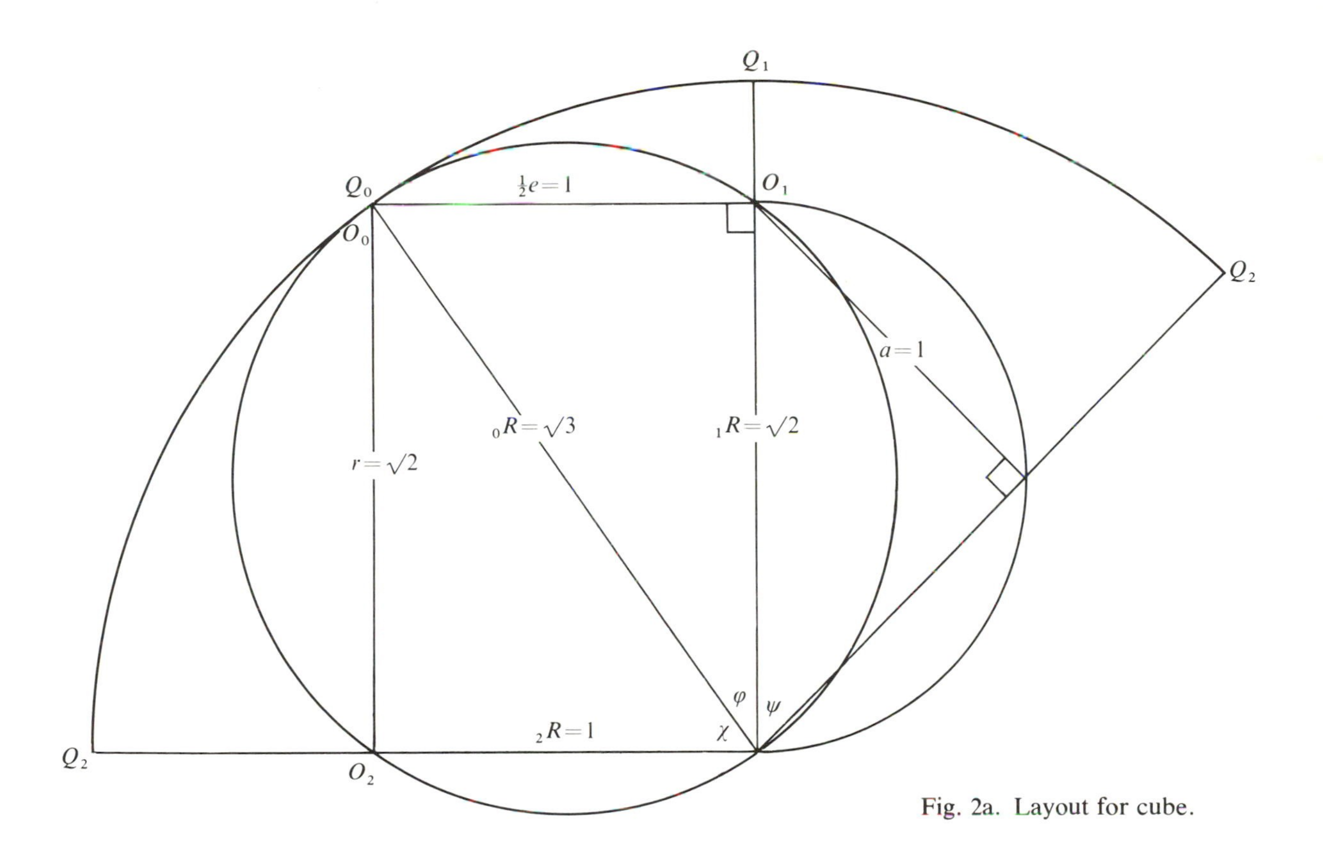

Fig. 2a. Layout for cube.

that a paper band can be designed for making a spherical model of the cube. Triangle $O_0O_1O_2$ is omitted, and Fig. 2a shows how to draw the layout of the orthoscheme and how the angles χ, ϕ, ψ, needed for the circular band, are derived from it.

The procedure is as follows: A circular arc of radius ${}_0R = O_0O_3 = \sqrt{3}$ is drawn first. The measure $\sqrt{3} = 1.732$ approximately. Note that ${}_0R$ is the radius of the circumsphere. Next draw a circle with O_0O_3 as diameter. Now open the compass to 1 unit of length and, with O_0 as center, mark the point O_1 on this circle. Draw O_3O_1 and produce it to Q_1 on the circular arc. The arc Q_0Q_1 is now one side of the spherical triangle $Q_0Q_1Q_2$. Notice that the angle at O_1 is a right angle, since it is inscribed in a semicircle.

The same theorem from plane geometry is now used to locate the point O_2, whose projection is Q_2, as shown on the left in Fig. 2a. Notice that $O_0O_2 = \sqrt{2}$, since it is half the diagonal of the square that is a face of the cube. This value is derived from the well-known theorem of Pythagoras. In fact $\sqrt{3}$ is itself derived from the same theorem, given that the edge length of the cube is 2 units.

So now open the compass to $\sqrt{2} = 1.414$ approximately and, with O_0 as center, mark the point O_2. Then O_3O_2 produced gives Q_2, which is the projection of the point O_2 onto the surface of the sphere. The arc Q_0Q_2 thus becomes the second side of the spherical triangle $Q_0Q_1Q_2$.

Line O_0O_2 is named r because it is the radius of the small circle circumscribing a face of the cube.

The third side of the spherical triangle $Q_0Q_1Q_2$ now follows as shown on the right in Fig. 2a. A semicircle must be drawn with O_1O_3 as diameter, so that another right angle may be inscribed in it. O_1O_2 is 1 unit of length, as may be seen from Fig. 1. Open the compass to this length and with O_1 as center, mark the point O_2 and project it to obtain Q_2. Arc Q_1Q_2 is the third side of the spherical triangle $Q_0Q_1Q_2$.

Line O_1O_2 is named a because it is an apothem, namely a line drawn from the incenter of a face to a midpoint of an edge. Since regular polyhedrons all have regular polygons for faces, the apothem is always a perpendicular bisector of an edge.

The angle measures associated with the spherical triangle $Q_0Q_1Q_2$ can now easily be determined by ordinary trigonometry from the respective right-angled triangles. Thus $\sin \phi = 1/\sqrt{3}$, $\sin \chi = \sqrt{2}/\sqrt{3}$, and $\sin \psi = 1/\sqrt{2}$. Hence the arc length $Q_0Q_1 = 35.264°$, the arc length $Q_0Q_2 = 54.736°$, and the arc length $Q_1Q_2 = 45°$.

Consideration of linear measures may now be dropped and the circular band of paper may be made using any convenient radius. This band is to be used as a template to multiply parts. A recommended linear or chord length of about 1 in. or 2.54 cm between radial lines is suggested and a width of about three-eighths of an inch or 1 cm for the band. Actual model making shows this gives best results with paper. The layout of the circular band for the spherical cube is shown in Fig. 2b.

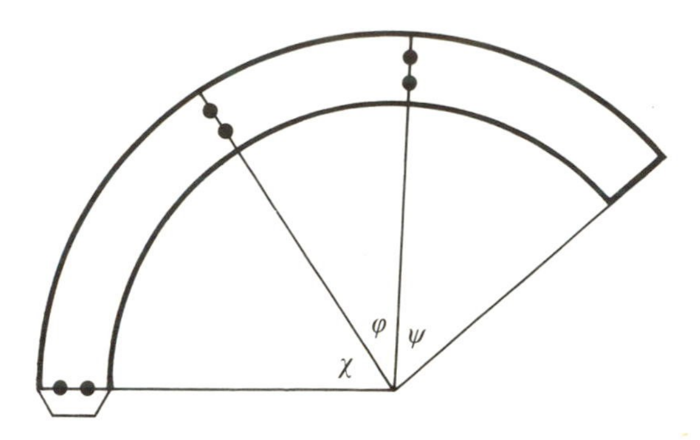

Fig. 2b. Circular band for cube.

To multiply the parts the following procedure may be used. Place the template over three, four, or five sheets of card or tag that have been stapled together. Outline the template by pencil and prick through the card at the radial lines and through the points marked by heavy dots in Fig. 2b. Now cut with scissors along the curved pencil outline as carefully as you can without allowing the paper to move. No scoring is needed because the paper folds easily along the pricked holes in the band. The parts are folded up for right-handed parts, and down for left-handed parts, or vice versa, whichever you decide on. A tab is left at one end of each band. This is used to glue the bands so as to form small spherical triangles.For the spherical cube, eight of these small spherical triangles are joined together, band to band, so that the tab joints disappear between the bands.

What emerges is a spherical dome whose edges form a spherical square. Six such domes are needed to complete the model of a spherical cube. Beyond all doubt before you complete the model you will see it coming out as a spherical octahedron as well, the dual with respect to the cube. The same feature appears in all the other regular and semiregular spherical models. In this respect the dual is most simply described as the polyhedron generated by great circle arcs joining the incenters of the faces of a given spherical polyhedron. In other words, here the polyhedron and its dual are both projected onto the surface of the same circumscribing sphere.

General instructions for making models

A word of caution is in place here for the dedicated model maker. The templates must be very accurately drawn to achieve even moderately good results. If in each case you make your own drawings, you will learn first of all how careful you must be about every detail. Then if things do not turn out well, you have only yourself to blame. It is a good idea to check out linear and angular measures against each other, so that calculation will always verify the correctness of geometrical drawing and vice versa. Even so you may find that slight adjustments may be needed as radial lines are folded. Make only a few parts at first, then start assembling them to see how they fit before you continue with the rest. Only experience can teach you what adjustments are best made to achieve desired results.

Do not be dismayed if at first the parts seem to fit rather poorly. Slight differences tend to disappear or compensate each other at vertex points in the completed models. You may find that these vertex points are never perfectly closed. You can observe this on the models shown in the photographs. But as the work moves toward completion the various parts begin to exert their structural tensions relative to each other. Somehow the end result always seems satisfactory, where misgivings or doubts may have previously tempted you to abandon the work. Remember that no model can be perfect. The scale at which the work is done as well as the material used has its limitations. A model is a model, and its purpose is to serve insight and understanding. From this point of view the perfect model can exist only in your mind. Or if you want to espouse the mystical approach of Johann Kepler, you would say it exists only in the mind of God the Creator.

The spherical octahedron

Having completed the spherical cube you can satisfy yourself that it is the spherical

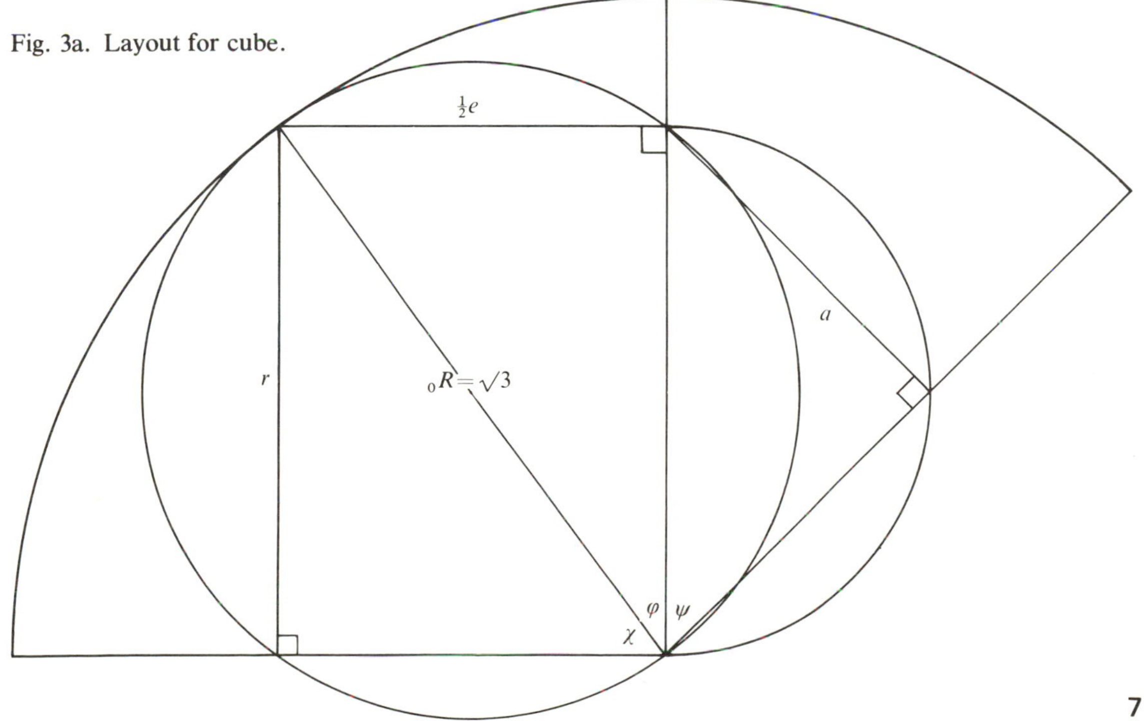

Fig. 3a. Layout for cube.

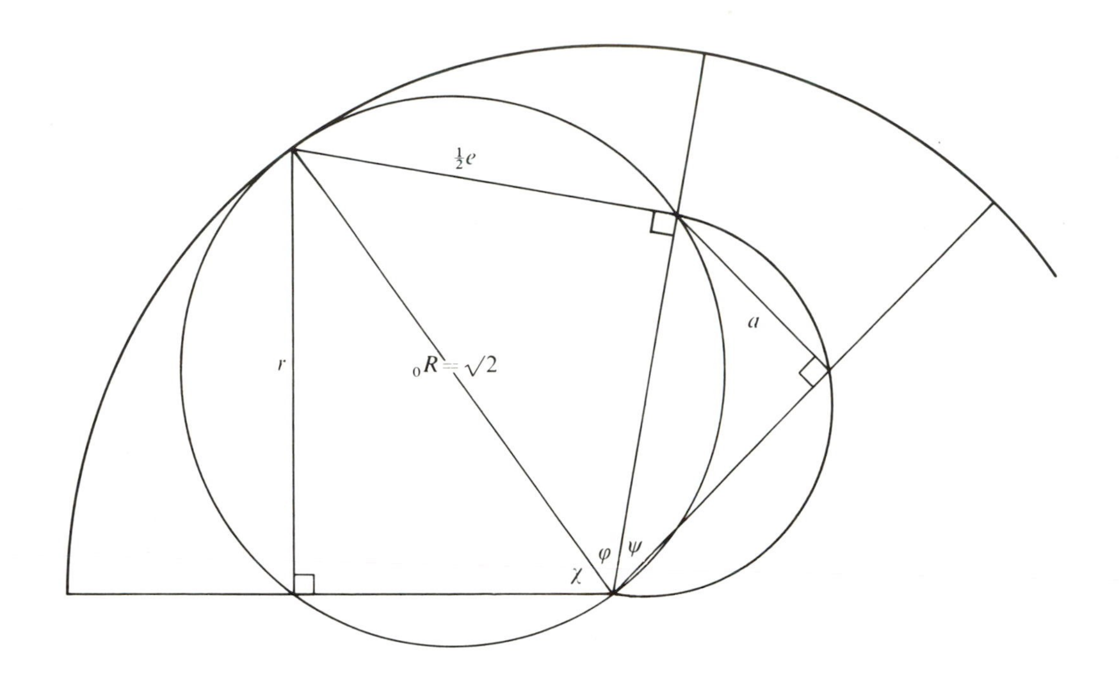

Fig. 3b. Layout for octahedron.

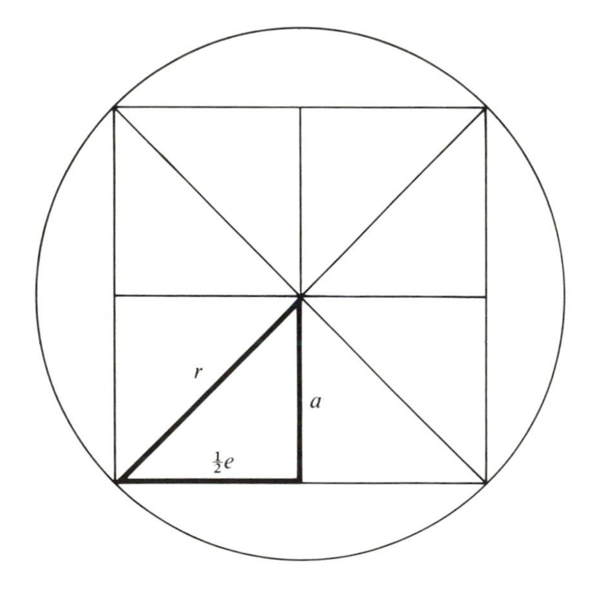

Fig. 3c. Facial plane for cube.

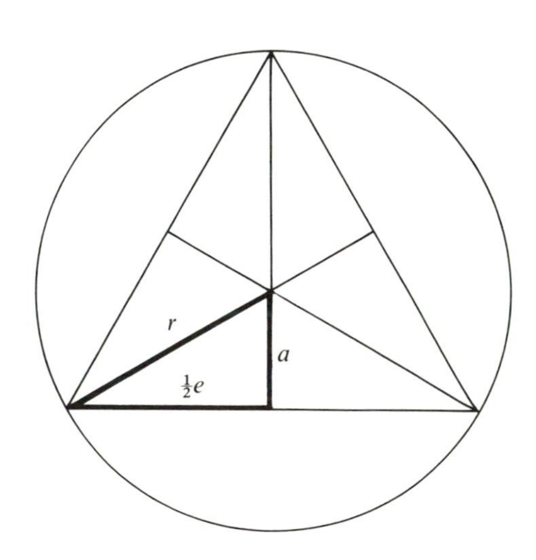

Fig. 3d. Facial plane for octahedron.

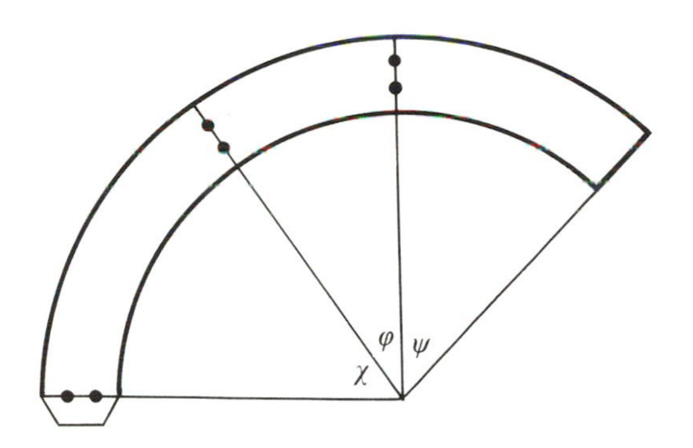

Fig. 3e. Circular band for cube.

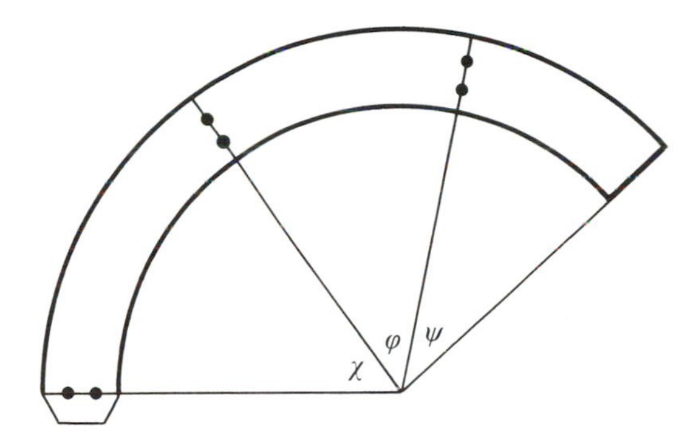

Fig. 3f. Circular band for octahedron.

octahedron as well, not only by looking at the model, but by going through the process of drawing the layout (Fig. 3b). The circular arc for the octahedron must have a radius of $\sqrt{2}$. This is easy to see if you notice that the octahedron has a set of four edges forming a square that lies on a plane of symmetry through the center of the polyhedron. Obviously the distance from one of these vertices to the center of the polyhedron must be half the diagonal of this square. If the edge length $e = 2$, the diagonal is $2\sqrt{2}$, thus making the radius $\sqrt{2}$. You see now that the only way in which the circular band for the spherical cube differs from that of the spherical octahedron, once you drop linear measurements and compare only the arc lengths, is in an interchange in the position of the central angles ϕ and ψ, whereas χ remains the same. Plates 1 and 2 show heavy lines to give emphasis to the arcs belonging to the spherical cube and the spherical octahedron, respectively. If you make only one model your mind must supply the emphasis, or to put this another way, your eye sees first one shape, then the other, depending on what you want to see.

The spherical tetrahedron

The spherical tetrahedron is illustrated in Plate 3. The tetrahedron, as you may know, is the simplest of all uniform polyhedrons, in fact of polyhedrons of any

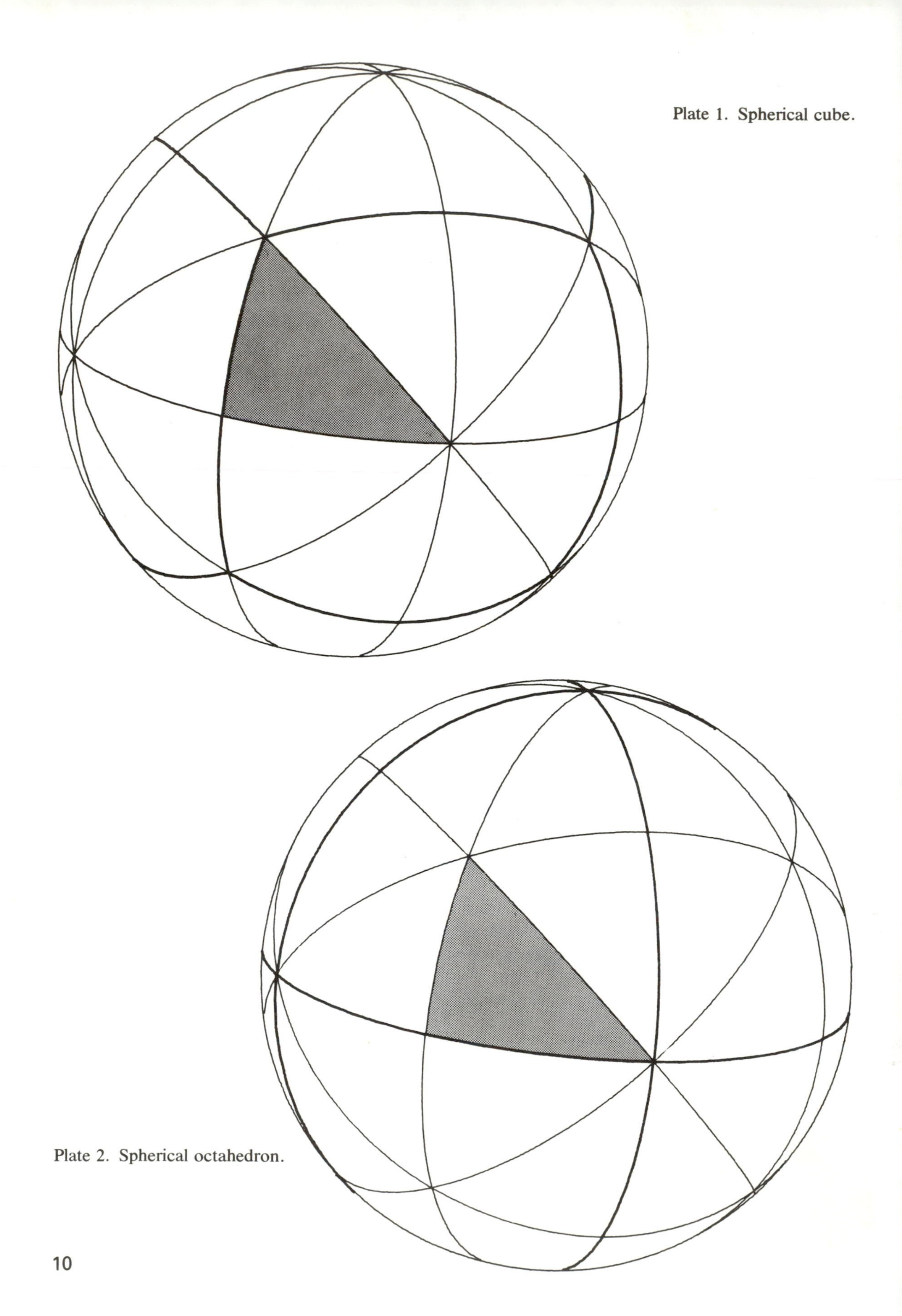

Plate 1. Spherical cube.

Plate 2. Spherical octahedron.

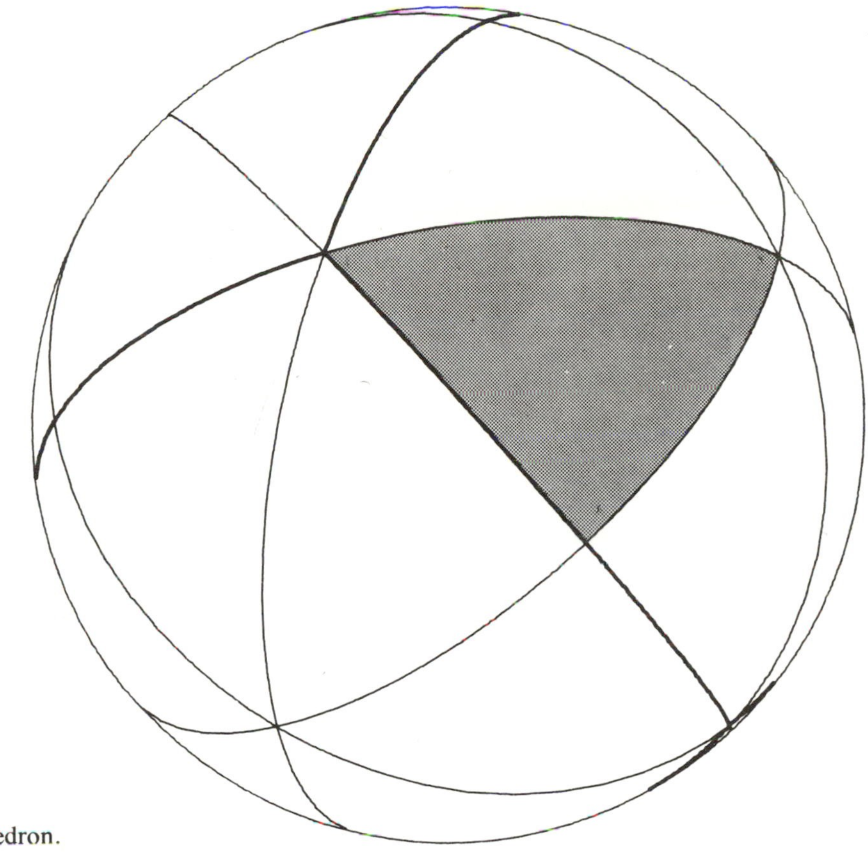

Plate 3. Spherical tetrahedron.

kind, having only four faces, six edges, and four vertices. It is unique in another way. It is the only uniform polyhedron that does not have vertices in diametrically opposite pairs. This means that its planes of symmetry each cut through only one vertex. The spherical model quickly shows that its dual is simply another spherical tetrahedron. Figure 4a gives the layout needed for deriving the circular band in this case; the band is shown in Fig. 4b.

For the tetrahedron ${}_0R$ = 1.225 approximately. This is the radius of the circumscribing sphere. Here and in all subsequent spherical models the radius of the circumscribing sphere will be given. In almost every instance it is a rather tricky affair to derive this radius in terms of an edge length. Euclid does a very elaborate job on this in the Thirteenth Book of his classic work *The elements*. If you are interested in the mathematical derivations, you may consult the more modern works listed in the References at the end of this book.

For the spherical tetrahedron it is best to keep the scale small, so that the paper bands will have less tendecy to become bowed or bent. The very simplicity of this model perhaps keeps if from being very attractive, or at least it seems to be less

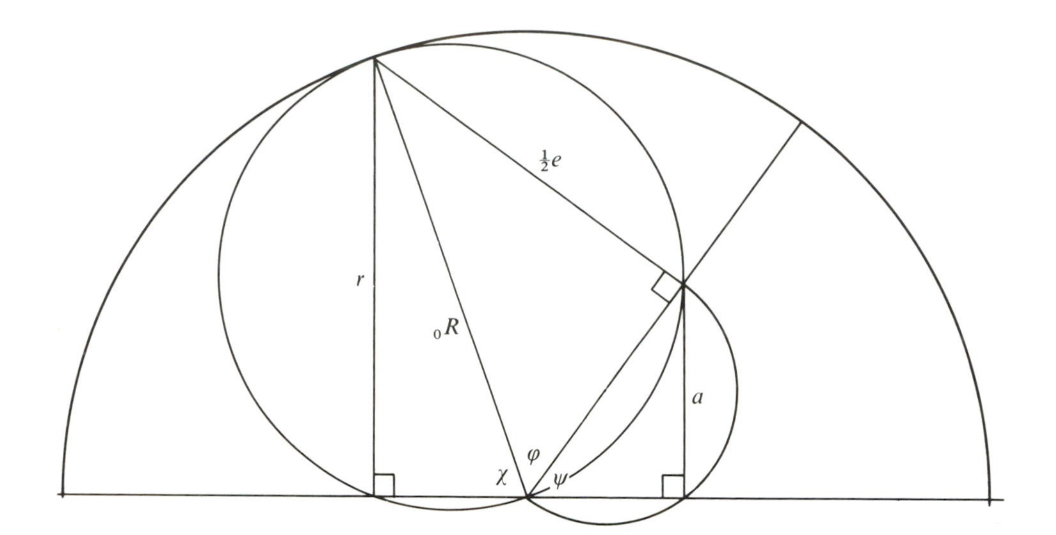

Fig. 4a. Layout for tetrahedron.

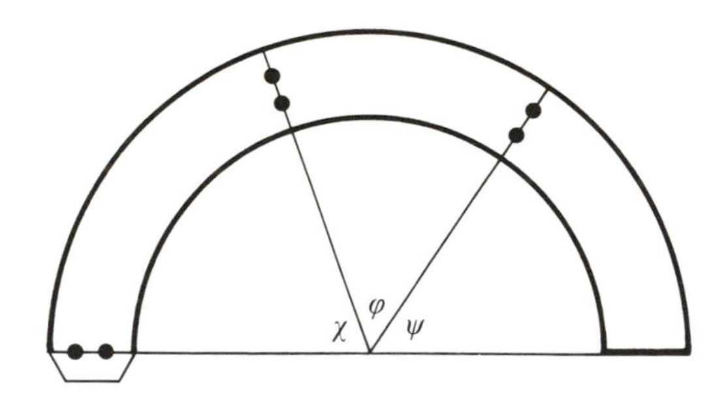

Fig. 4b. Circular band for tetrahedron.

attractive than the other spherical models. Beauty usually demands some complexity, because it arises from the harmonious relationship of parts. But it is a good model to have on hand for inspection; further consideration is given to it later on in the book.

The spherical icosahedron and dodecahedron

The spherical icosahedron is the third of the regular spherical models. It turns out to be the most attractive, perhaps because the so-called golden ratio τ appears in its linear measurements. The same beauty manifests itself in all the polyhedrons that belong to the icosahedral symmetry group. The dual of the icosahedron is the dodecahedron. Here too, as in the case of the cube and the octahedron, one model serves for both. Look again at Photo 4-5 and observe this.

You can make this model by using either of the layouts shown in Fig. 5a or 5b. If you assemble triangle faces, you must arrange them as the faces of an

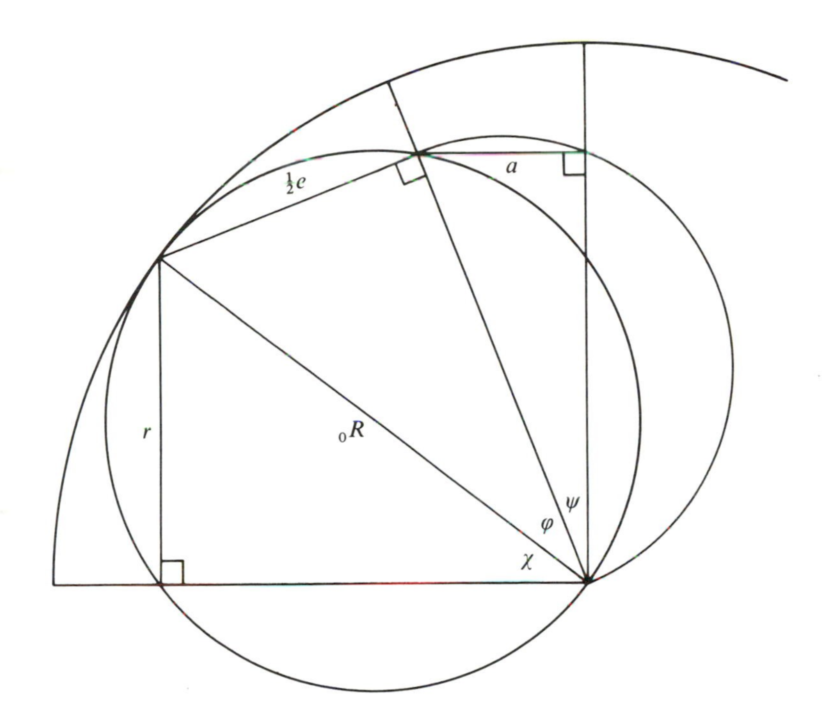

Fig. 5a. Layout for icosahedron.

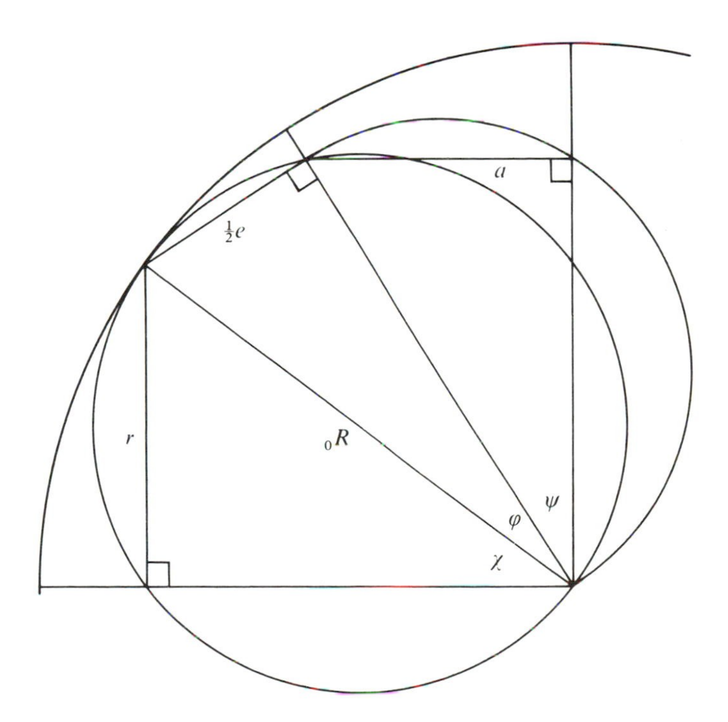

Fig. 5b. Layout for dodecahedron.

Fig. 5c. Facial plane for dodecahedron.

icosahedron. If you assemble pentagon faces, it is the dodecahedral arrangement that is needed. In practice the latter seems preferable. For triangle faces, r and a have already been calculated. You can find r and a for the pentagon by using ordinary trigonometry applied to Fig. 5c. In comparing the two layouts you see that ϕ and ψ are interchanged, whereas χ remains the same – this is the same situation as was found previously for the cube and the octahedron. The drawings of Plates 4 and 5 show heavy lines to emphasize the spherical icosahedron and the spherical dodecahedron. The bands are shown in Figs. 5d and 5e.

The beauty of these models will be greatly enhanced by an appropriate use of color. If the paper used is colored tag, all the left-handed parts may be white, and all the right-handed ones colored. Color effects can be used in all the spherical models of this book. It will be left for you to work out whatever you may choose. No further references will be made to color in what follows.

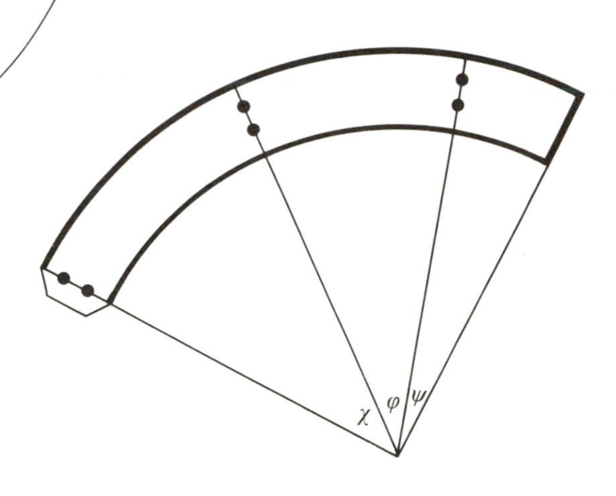

Fig. 5d. Circular band for icosahedron.

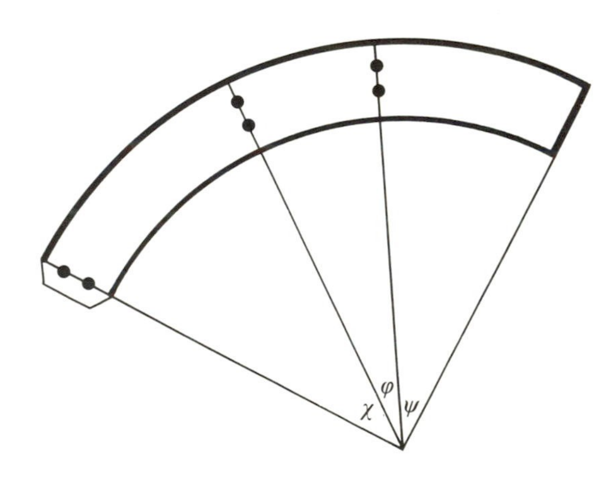

Fig. 5e. Circular band for dodecahedron.

Plate 4. Spherical icosahedron.

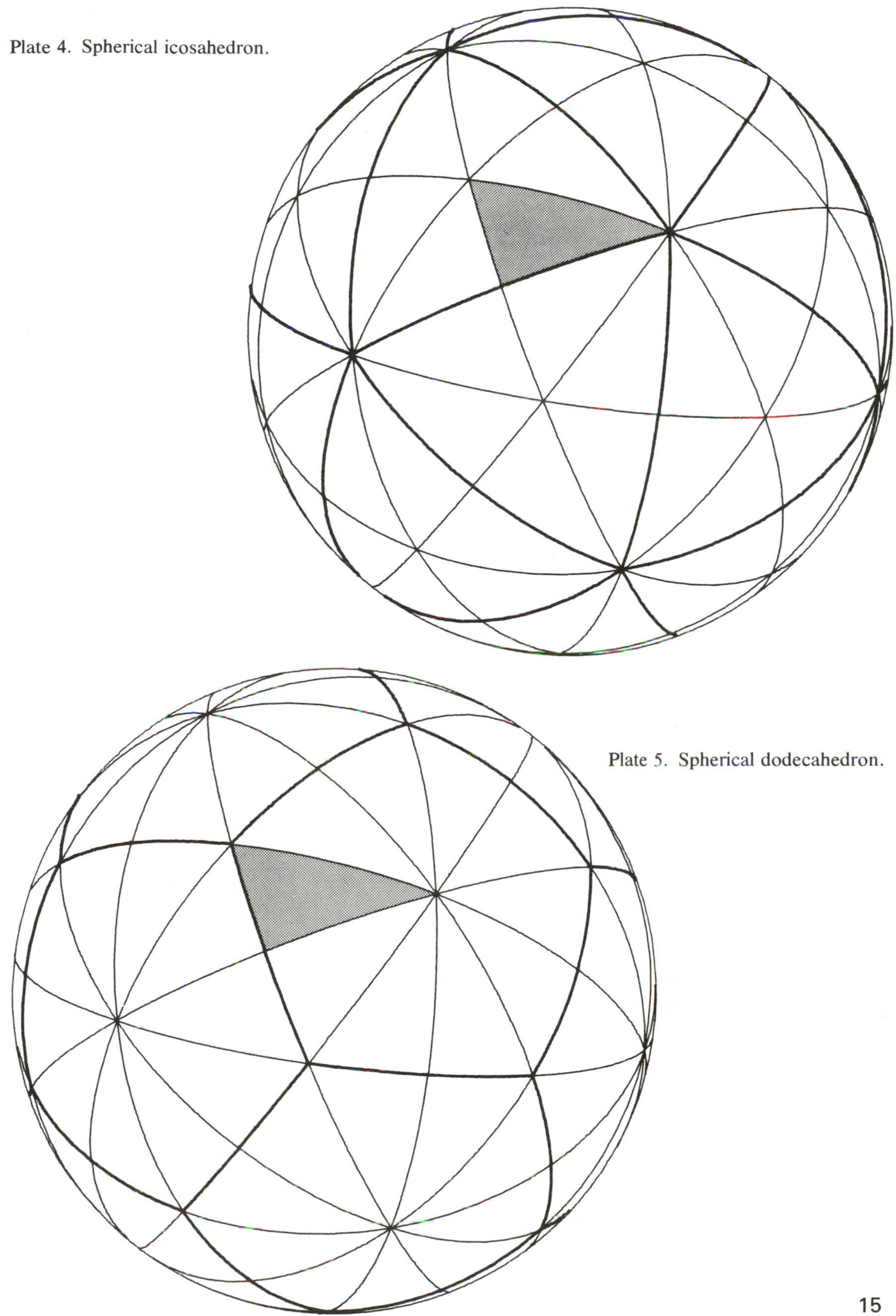

Plate 5. Spherical dodecahedron.

The polyhedral kaleidoscope

If you have now made the three regular spherical models, you are in a position to use them for attaining a better understanding of the polyhedral kaledioscope. The basic idea here involves reflection of a point in a line. A point P' is said to be the reflection or image of a point P in a line l when the line segment PP' is perpendicular to l and the distances of P from l and of P' from l are equal; that is, line l is the perpendicular bisector of line segment PP' (see Fig. 6).

A plane polygon can be thought of as being generated by repeated reflections of a given point P in a set of lines all radiating from a fixed point O, the lines having suitable angular distances between them. Reflection may be replaced by the idea of rotation. Reflection and rotation are types of mathematical transformations. To simplify matters the discussion here will limit itself to mathematical reflection.

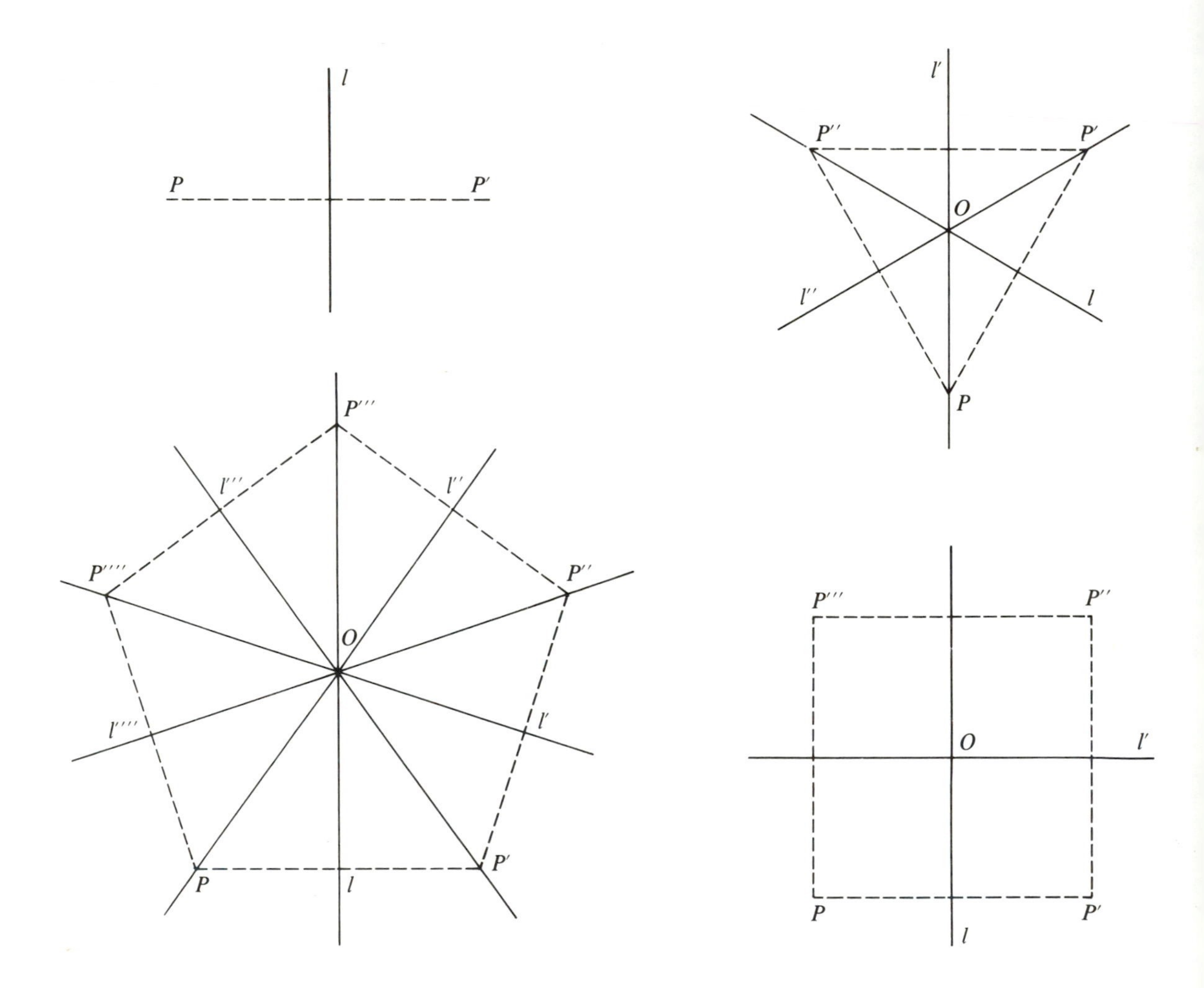

Fig. 6. Reflections of a point P in a line l.

Figure 6 illustrates the idea of reflection applied to an equilateral triangle, a square, and a pentagon.

If two mirrors are used and set upright on a desk so that the angle between them is 120°, and if a small object is used to mark a point and this object is placed midway between the mirrors, the object with its images marks the vertices of an equilateral triangle. If the angle between the mirrors is changed to 90°, the result is a square. If the angle is 72°, the result is a pentagon. This arrangement of mirrors is called a kaleidoscope.

The idea of a point and its images can now be applied very effectively to a spherical surface, where the point is any given point on the spherical surface and where the lines are geodesics or great circle arcs.

Three mirrors are needed for a polyhedral kaleidoscope. They must be cut as circle sectors whose central angles are precisely the values of χ, ϕ, ψ already calculated for making the spherical models in paper. You can save yourself a lot of trouble by forgetting about the use of mirrors, except possibly to try them and see for yourself that the results are not completely satisfactory. The paper models you now have on hand should give you the same results in a far more satisfactory way.

Examine the tetrahedral model first. Take as the given point P one of the points where three great circles intersect. Actually there are eight such points, but fix your attention on one alone. By reflection in one of the geodesics intersecting at O, call this geodesic line l, the point P gives rise to its image point P'. The point P' reflected in the geodesic line l' generates the point P''. Finally the reflection of P'' in l'' returns you to P (see Fig. 7). The fourth vertex of the tetrahedron is found diametrically opposite to O. It is the reflection of any one of the other three points: P', P'', P'''.

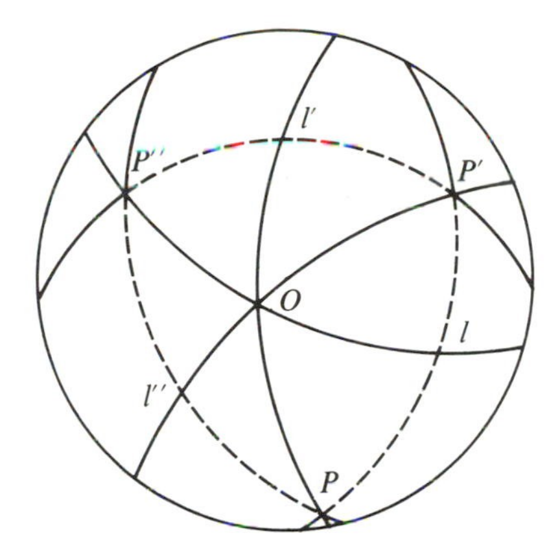

Fig. 7. Reflections generating a spherical tetrahedron.

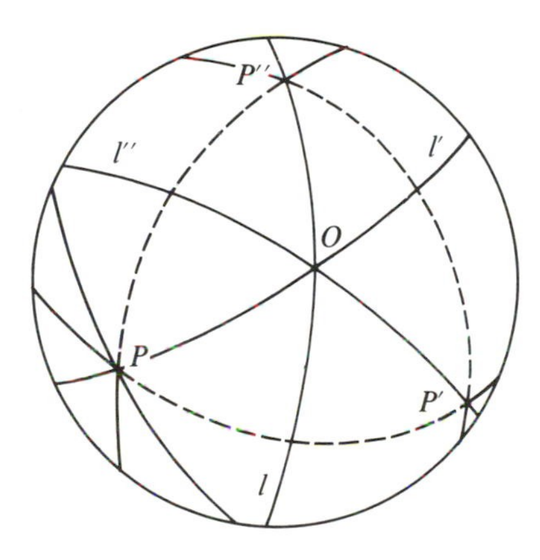

Fig. 8a. Reflections generating a spherical octahedron.

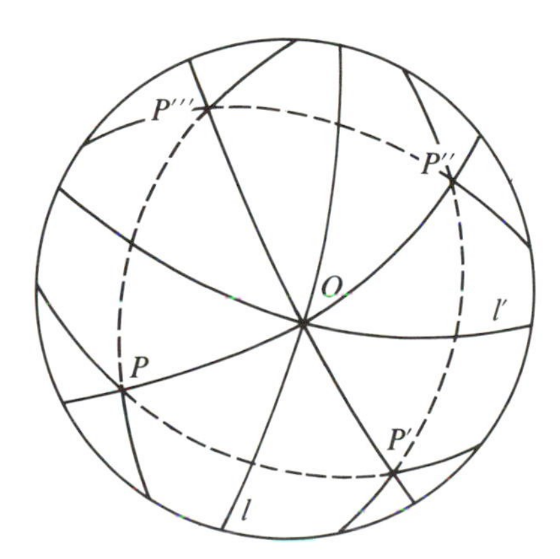

Fig. 8b. Reflections generating a spherical cube.

Look next at the octahedral model. If you fix your attention on a point that can be thought of as the incenter of a spherical square and name that point P, then its continuing reflections in the set of geodesics intersecting at O, the incenter of a spherical equilateral triangle, will generate the points marking the vertices of one face of the octehedron (see Figs. 8a and 8b). Continuing reflections over the whole surface of the sphere will generate the other vertices and hence the other faces of the octahedron.

If you now shift your attention to point O and rename it P, namely the point to be reflected, you will see that its continuing reflections over the whole surface of the sphere will generate the spherical cube. The role played by points P and O is thus seen to be interchangeable, and this illustrates the dual relationship of the octahedron and the cube.

Finally, in examining the icosahedral model you will see the same principles at work. If you fix your attention on the incenter of a spherical pentagon, its continuing reflections over the entire surface of the sphere will mark the vertices and hence the faces and edges of the spherical icosahedron. Reciprocally, an incenter of a spherical equilateral triangle will generate the dual polyhedron, the spherical dodecahedron. Figures 9a and 9b illustrate the situation for one face, respectively, of each.

You may by this time have noticed the existence of other points of intersection of the geodesics in the three spherical models. They are in each instance points at which the geodesics cross at right angles to each other. Can these points be chosen as the given point P? The answer is yes, they can, but the results that flow from such a choice will be seen in the next section. There other suitable points will also be chosen to generate all thirteen semiregular spherical polyhedrons.

Summary

Table 1 summarizes the results of this section. It supplies data that you will be using in the following sections of this book as well. Table 1 along with Table 2 should be studied in connection with Fig. 10, which shows a generalized layout for finding the central angles χ, ϕ, ψ, along with six kinds of regular polygons. Note that an edge length of 2 units is maintained for each

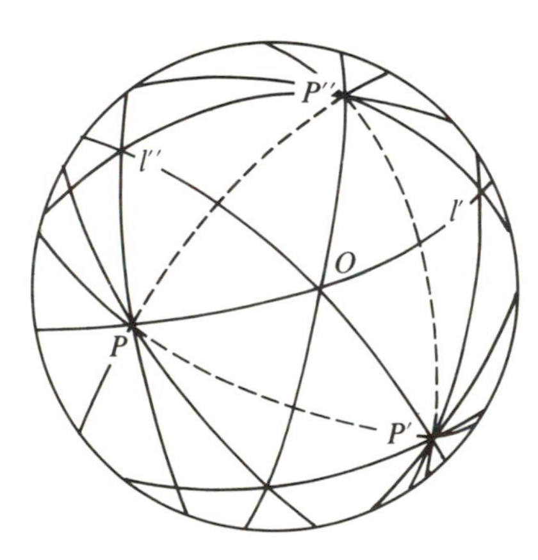

Fig. 9a. Reflections generating a spherical icosahedron.

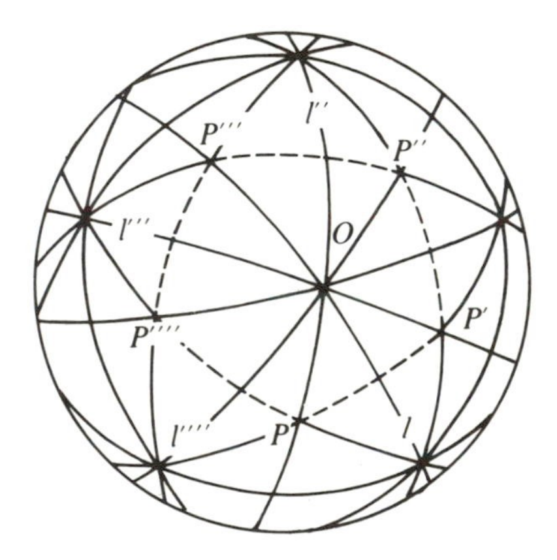

Fig. 9b Reflections generating a spherical dodecahedron.

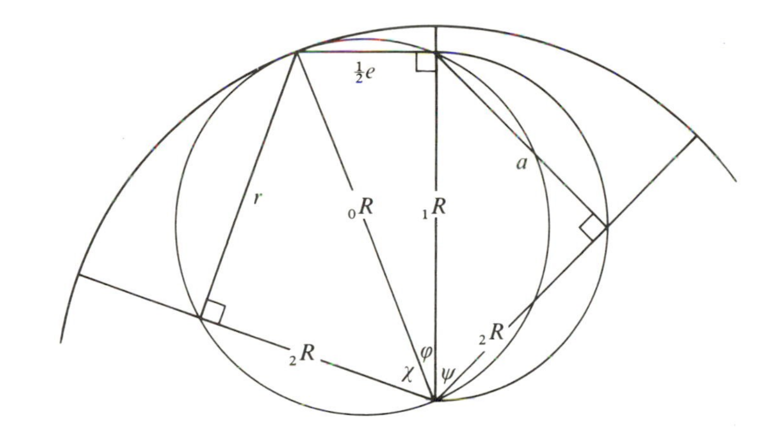

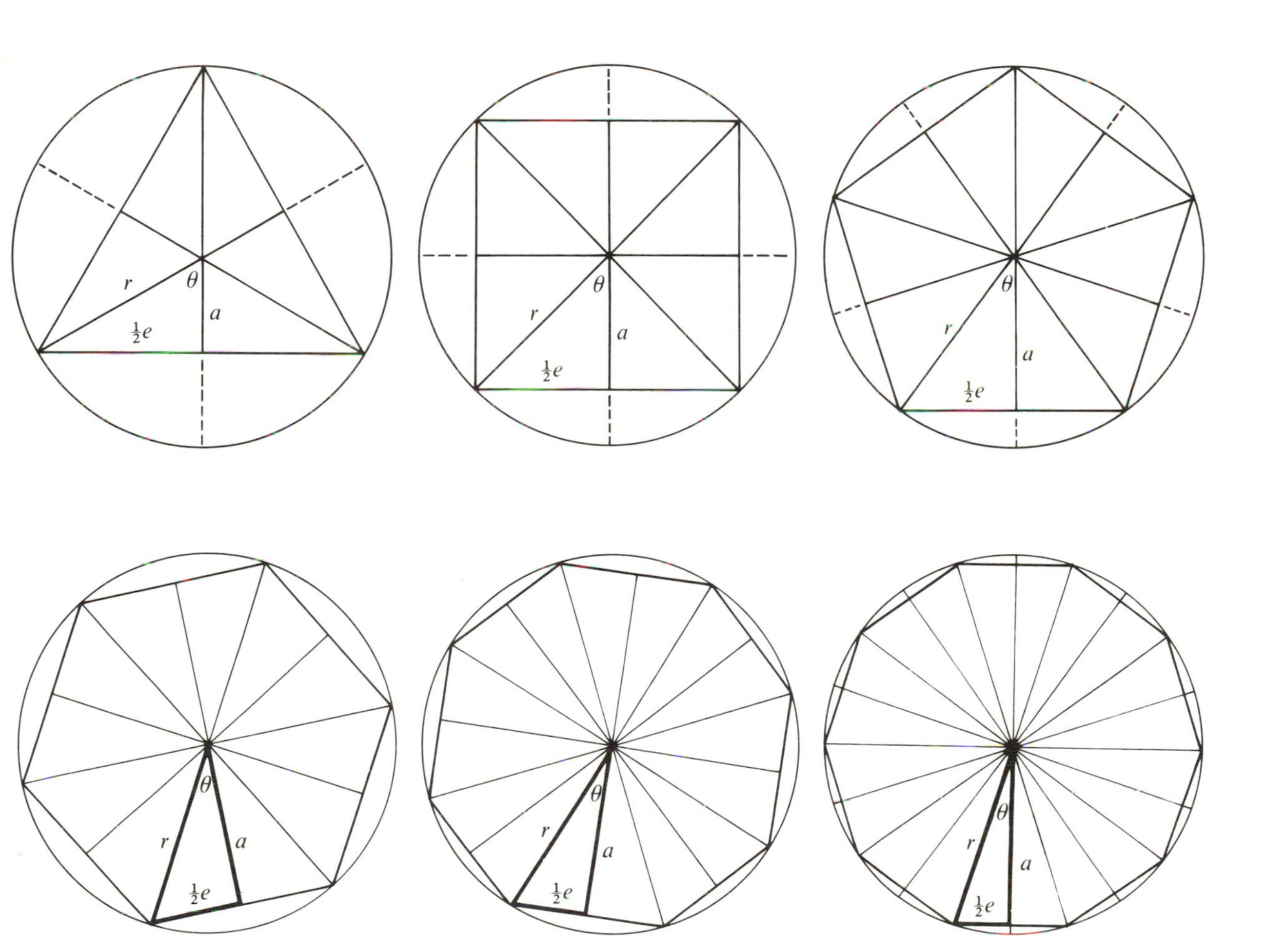

Fig. 10. A generalized layout with six kinds of polygons used as facial planes for polyhedrons.

case. Only the equilateral triangle, square, and pentagon occur in the regular solids. The hexagon, octagon, and decagon are given for future reference.

The data may be calculated by using the mathematical relationships given here. (You are invited to verify all these results. This need not be done all at once – it may precede or follow the making of each individual model.)

$$_0R = \sqrt{(_1R^2 + 1)}$$
$$_1R = \sqrt{(_0R^2 - 1)}$$
$$_2R = \sqrt{(_1R^2 - a^2)}$$
$$r^2 - a^2 = 1$$
$$e = 2$$
$$\tfrac{1}{2}e = 1$$

$$\cos\phi = \frac{_1R}{_0R}$$
$$\cos\chi = \frac{_2R}{_0R}$$
$$\cos\psi = \frac{_2R}{_1R}$$

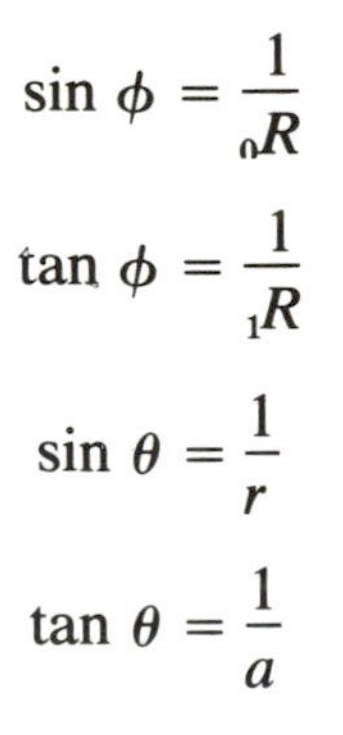

$$\sin\phi = \frac{1}{_0R}$$
$$\tan\phi = \frac{1}{_1R}$$
$$\sin\theta = \frac{1}{r}$$
$$\tan\theta = \frac{1}{a}$$

Fig. 10a. A right-angled triangle.

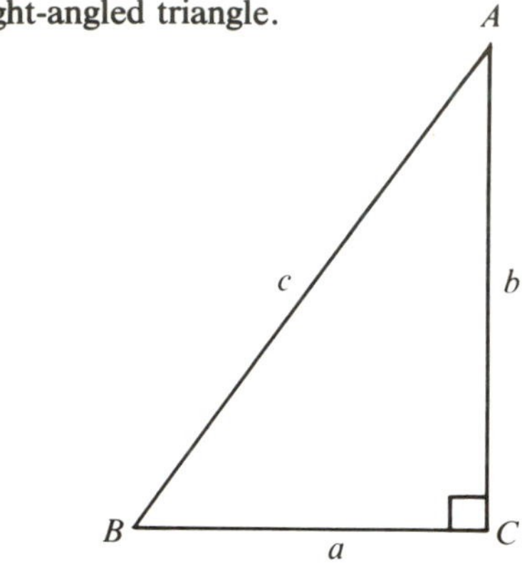

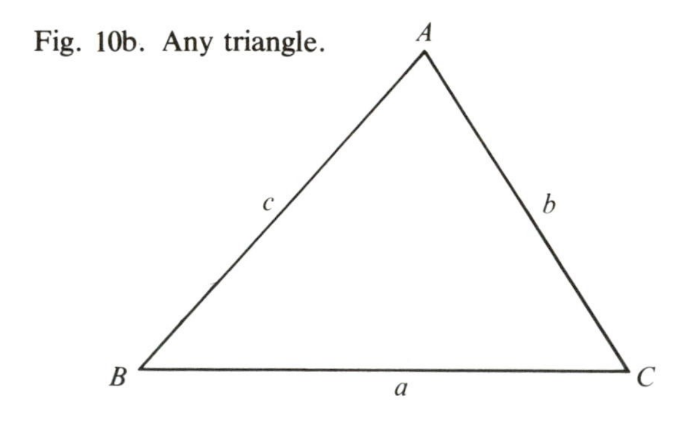

Fig. 10b. Any triangle.

Table 1. *The five regular solids*[a]

Polyhedron	$e = 2$	$_0R$	$_1R$	$_2R$	χ	ϕ	ψ
1 Tetrahedron	{3, 3}	1.225	0.707	0.408	70.528	54.735	54.735
2 Octahedron	{3, 4}	1.414	1.000	0.816	54.735	45.000	35.264
3 Hexahedron (cube)	{4, 3}	1.732	1.414	1.000	54.735	35.264	45.000
4 Icosahedron	{3, 5}	1.902	1.618	1.512	37.378	31.717	20.905
5 Dodecahedron	{5, 3}	2.803	2.618	2.227	37.378	20.905	31.717

[a]The linear measures $_0R$, $_1R$, and $_2R$ are in the same units of length as the edge length $e = 2$; the angular measures χ, ϕ, and ψ are in degrees.

Table 2. *Polygons used for regular and semiregular solids*[a]

Polygon	$e = 2$	r	a	θ
Triangle	{3}	1.155	0.577	60
Square	{4}	1.414	1.000	45
Pentagon	{5}	1.701	1.376	36
Hexagon	{6}	2.000	1.732	30
Octagon	{8}	2.613	2.414	22.5
Decagon	{10}	3.236	3.078	18

[a]The linear measures r and a are in the same units of length as the edge length $e = 2$; the angular measure θ is in degrees.

Some general formulas from ordinary trigonometry that you may find useful here as well as later are the following:

In any right-angled triangle (see Fig. 10a),

$$\sin A = \frac{a}{c} \qquad \cos A = \frac{b}{c} \qquad \tan A = \frac{a}{b}$$

For any triangle (see Fig. 10b),

$$\frac{\sin A}{a} = \frac{\sin B}{b} = \frac{\sin C}{c}$$

$$a^2 = b^2 + c^2 - 2bc \cos A$$

II. The semiregular spherical models

At the end of Section I the question was raised: What would be the result of continuing reflections over the entire surface of a sphere if a point *P* were chosen that is the point at which two geodesics intersect each other at right angles? To see what happens examine again the tetrahedral model. Figure 11 shows dotted lines connecting the midpoints *P*, *P'*, *P''* on the edges of a tetrahedral face. (The designation "spherical" will henceforth be understood in most cases without being expressed.) Point *P* is precisely the point at which two geodesics cross at right angles. Further observation shows that the complete set of reflected points, including *P* itself, comes to six. These mark the vertices of the regular octahedron (see Plate 6). No new polyhedron turns up in this case. Note, however, that the dotted lines of Fig. 11 are not actually present as paper bands in the model. Your imagination must supply them. If they were actually there, you would see that they belong to three more great circles, which, added to the six already present, will give a total of nine, precisely the number found in the octahedral model. So that is why no new polyhedron turns up.

This brings you next to examine the octahedral model. Figure 12 shows the situation: Point *P* is a point at which two geodesics intersect at right angles; the dotted lines connect point *P* to its reflections; around the incenter *O* you see a smaller equilateral triangle inside a larger one. But the point *P* can also be thought of as the midpoint of one of the edges of a cube. The dotted lines of Fig. 12 show the continuing reflections generating a smaller square inside a larger one, where *O'* is the incenter of such a square. This is easier to see on the model than it is from Fig. 12.

With the model in your hand, you can follow in your imagination the continuing reflections over the entire surface of the

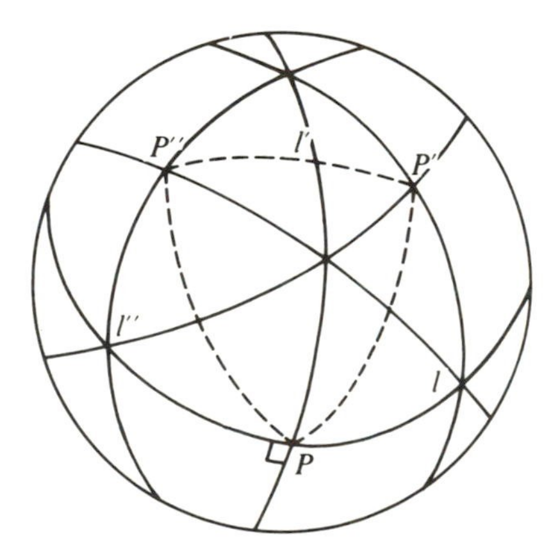

Fig. 11. Reflections on the surface of a tetrahedral model generating a spherical octahedron.

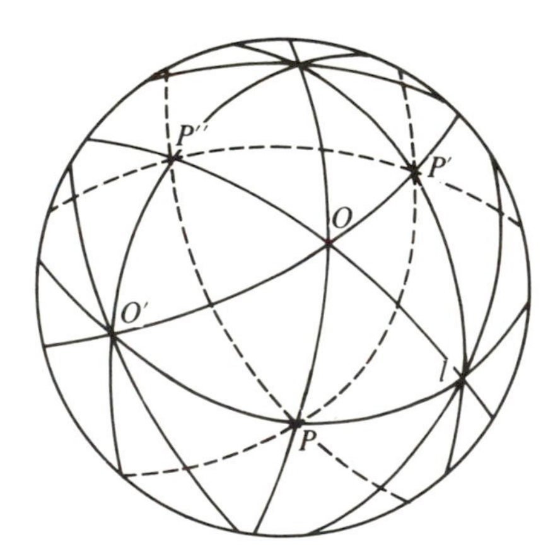

Fig. 12. Reflections on the surface of an octahedral model generating a spherical cuboctahedron.

Plate 6. Spherical octahedron.

Plate 7. Spherical cuboctahedron.

sphere and you may thus be able to recognize that the result is the cuboctahedron. In Plate 7 the shaded triangle is the same one as that shaded in Plates 1 and 2. This shows the relationship of the cuboctahedron to the octahedron and the cube. You realize, of course, that the dotted lines of Fig. 12 are not actually present in the octahedral model. You might be thinking: Wouldn't it be nice if these could actually be inserted as paper bands to get a model of the cuboctahedron that leaves nothing to the imagination? Fortunately this can be done.

It will be done, not by inserting paper bands in an octahedral model, but by generalizing the procedure already described for deriving paper bands for the regular spherical models.

The spherical cuboctahedron

Figure 13a shows the layout for the cuboctahedron. The first thing to be noted is that with the semiregular polyhedrons more than one kind of face is found in each polyhedron. In the cuboctahedron there are eight triangles and six squares. This means that two kinds of orthoschemes enter the picture. Hence you will need two different kinds of circular bands. But since both bands must have the same radius, ${}_0R$ = 2.000, which is the radius of the sphere circumscribing the cuboctahedron, only one circular arc is needed in the layout.

The construction is as follows: First the circular arc is drawn with center O_3 and with $O_3O_2 = {}_0R$ as radius. Then a circle is drawn of which O_3O_0 becomes the diameter. On this circle, to the right, O_1 is first marked off, with compass open to 1 unit of length and with O_0 as center. The use of subscripts in Fig. 13a should not be too difficult to interpret: O_0, O_1, O_2, O_3 are the points defined as in Fig. 1; so too, ${}_0R$, ${}_1R$, ${}_2R$. If a second subscript is used, it designates a relationship to a polygon of 3, 4, 5, 6, 8, 10 sides. The subscripts used with the letters a and r specify the polygon whose apothem and radius are needed.

Returning now to the layout, $O_{2,3}$ and $O_{2,4}$ are marked with compass open to r_3 and r_4, with O_0 as center. Both r_3 and r_4 are in the same semicircle so that they are at right angles to the radial lines $O_3O_{2,3}$ and $O_3O_{2,4}$. After the semicircle is drawn with O_3O_1 as diameter, a_3 and a_4 can be marked off on it, to the right, with O_1 as the center for the compass. The measures χ_4, χ_3, ϕ, ψ_3, ψ_4 now follow easily from the respective right-angled triangles, using the formulas given at the end of Section I. The circular bands can then be made as shown in Figs. 13b and 13c.

To assemble the spherical cuboctahedron make eight bands that belong to the square and complete the spherical dome having a spherical square for its edges. Make six bands that belong to the triangle and complete this in a similar fashion. Continue to make more of each and cement the squares and triangles together, band to band, following the arrangement of faces that pertains to the cuboctahedron. Photo 11 (see later in this section) shows the completed model.

By examining the completed model of the cuboctahedron, you will see that χ_3 is the same arc length as ψ of the octahedron and as ϕ of the cube, whereas $\psi_3 + \psi_4 = \chi$ of the octahedron or of the cube. In this way the relationship of these polyhedrons as duals is made explicit. At the same time the overall effect is very attractive.

The spherical icosidodecahedron

Turning now to the icosahedral model, you will find that the icosidodecahedron is generated from it in much the same way as

Fig. 13a. Layout for cuboctahedron.

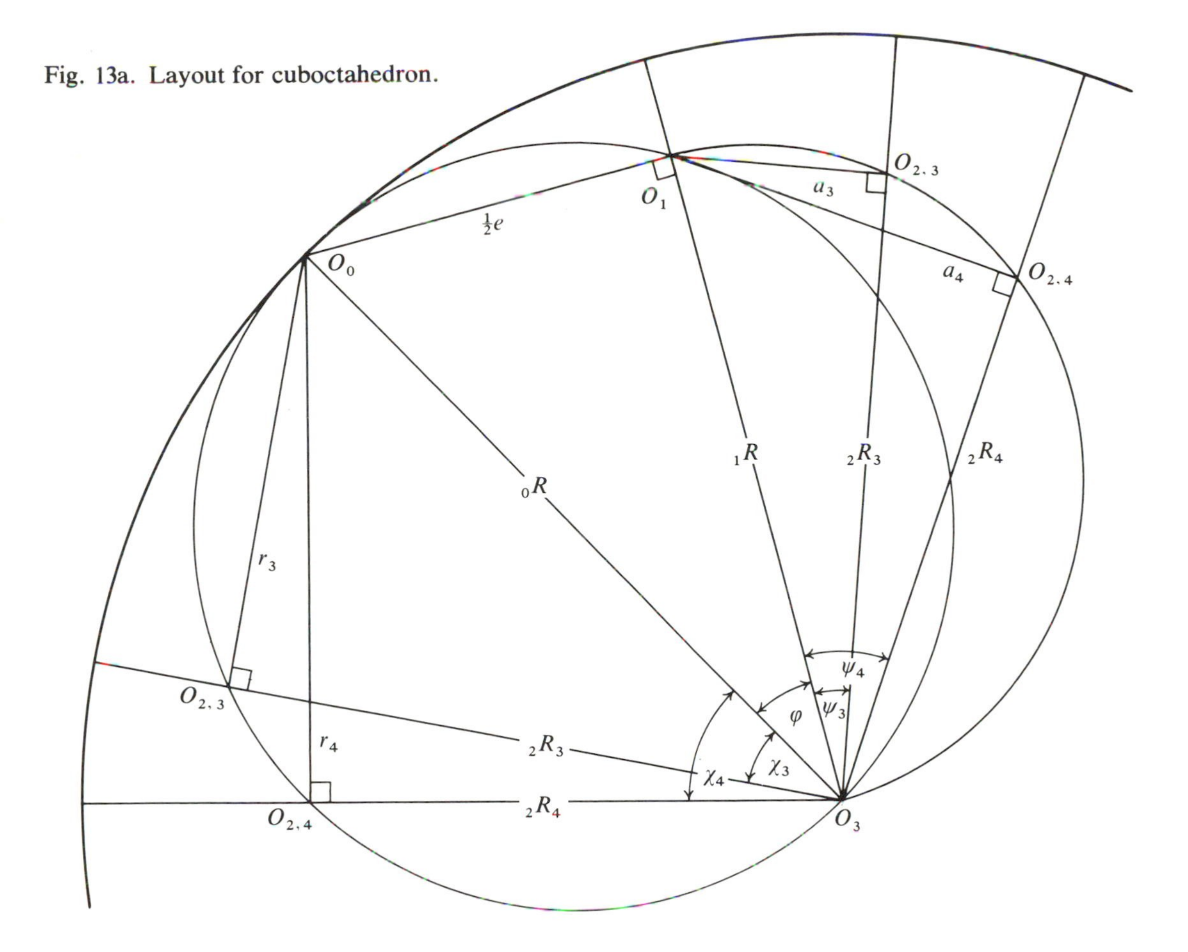

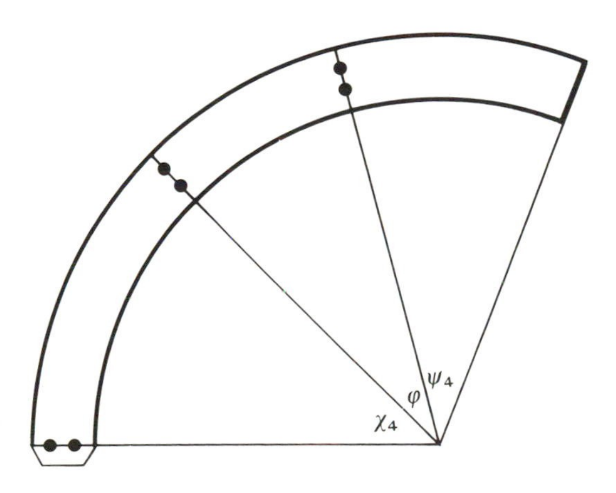

Fig. 13b. Band for square face.

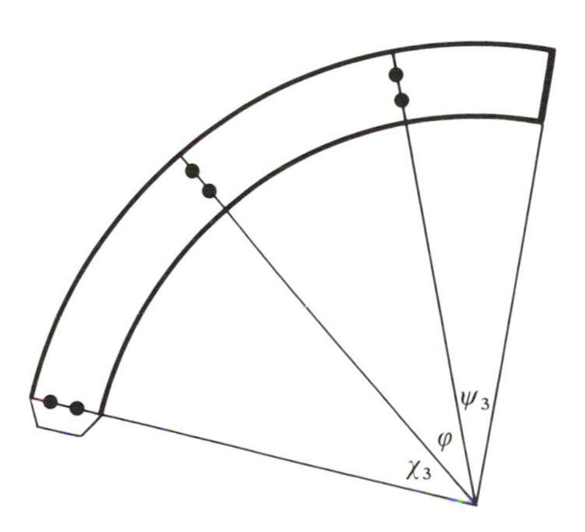

Fig. 13c. Band for triangle face.

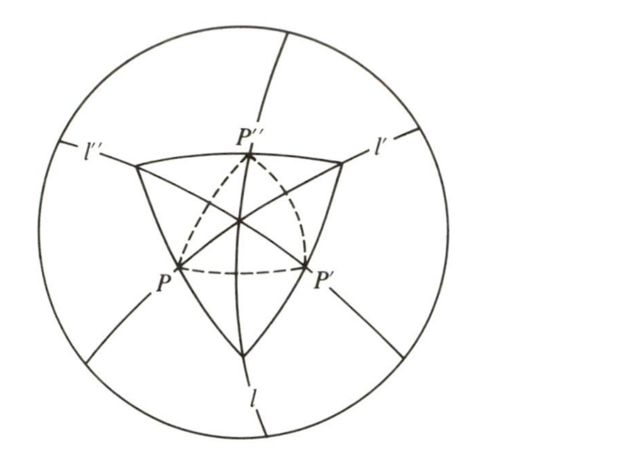

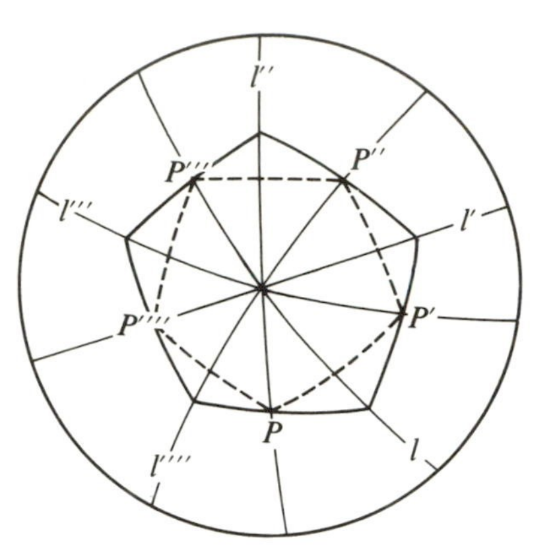

Fig. 14. Reflections on the surface of an icosahedral model generating a spherical icosidodecahedron.

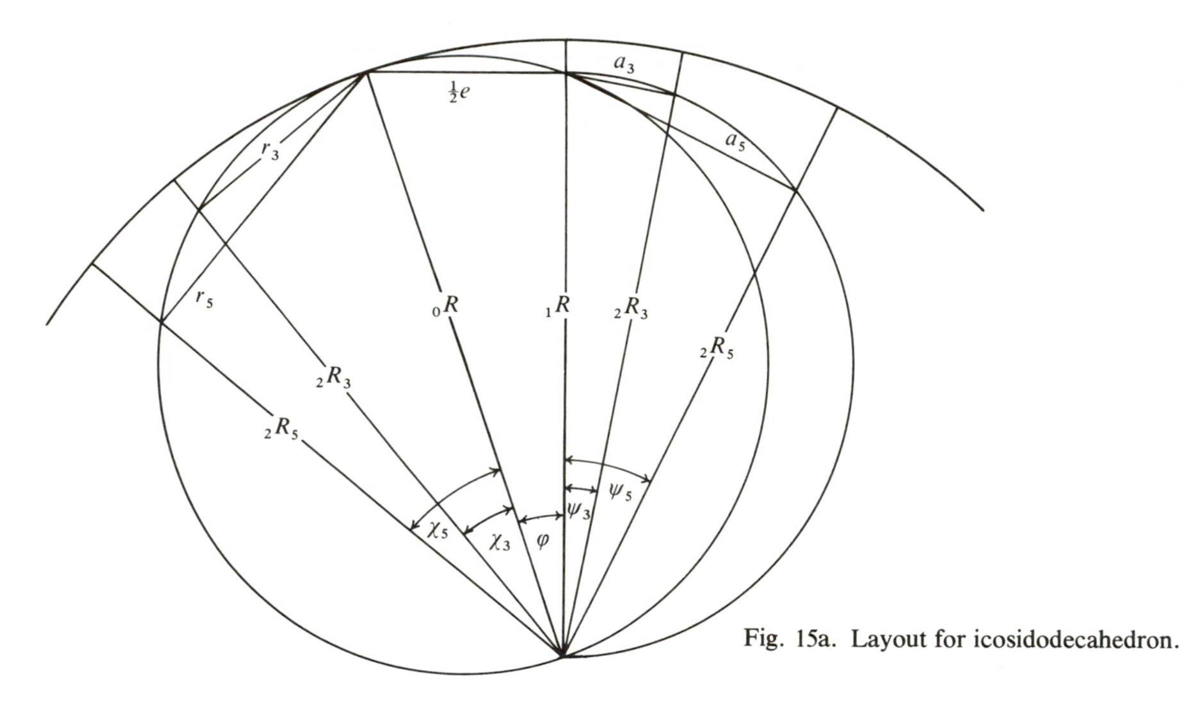

Fig. 15a. Layout for icosidodecahedron.

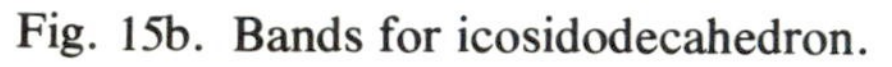

Fig. 15b. Bands for icosidodecahedron.

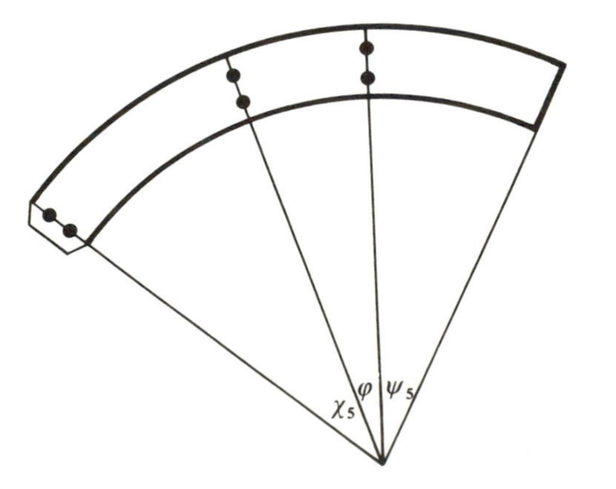

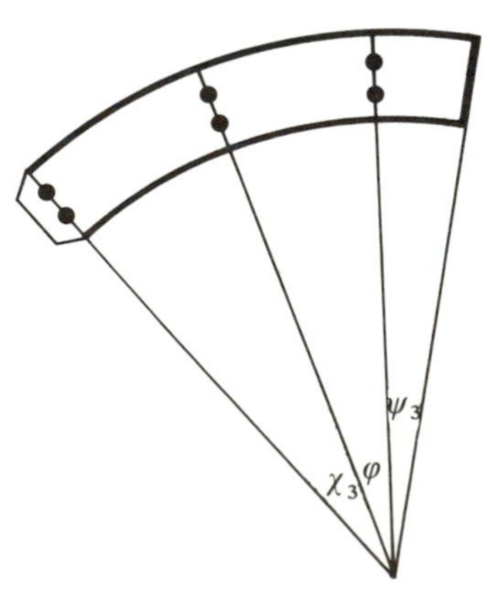

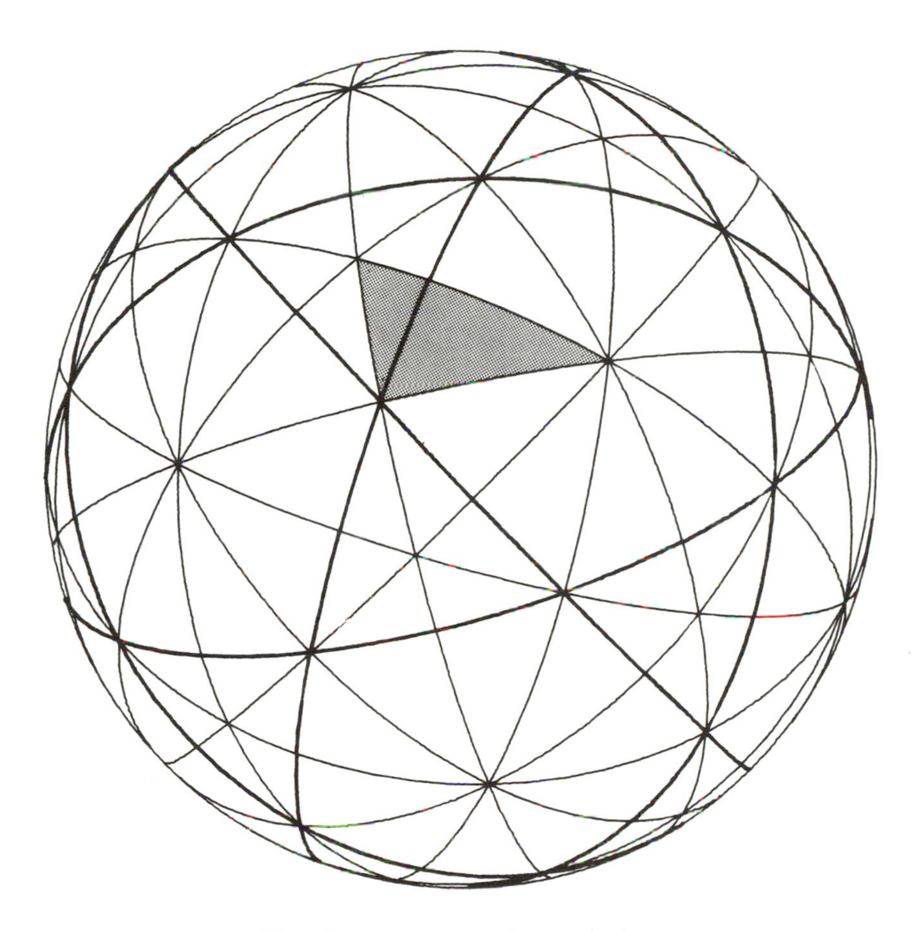

Plate 8. Spherical icosidodecahedron.

the cuboctahedron was generated from the octahedral model. Figure 14 shows a simplified version of the situation: Point *P* is the point at which two geodesics cross at right angles; the dotted lines show a smaller triangle and a smaller pentagon inside the larger ones. The same point *P* is on an edge of the triangle or of the pentagon, depending on which you choose to see. Thus, by using the icosahedral model, you can follow in your imagination the results of these reflections over the entire surface of the sphere.

However, you will find it more satisfactory to make the spherical icosidodecahedron with all its right spherical triangles clearly visible. The layout is given in Fig. 15a, and the bands in Fig. 15b. You should have no trouble interpreting these and calculating the angular measures needed for the bands. Once it is finished you will see how it is explicitly related to the icosahedral model. See Plate 8.

Notice that, although the radius of the circumsphere ${}_0R = 3.236$, the circular bands may be reduced in size. Radii of 2–4 in. or about 5–10 cm are suitable. Photo 12 and subsequent photos show models in this range of sizes.

Spherical triangles as characteristic triangles

Before proceeding further it will be useful to study carefully the following elaboration. Plates 9–11 show the three regular spherical models with one triangle shaded in each case. This triangle is called a characteristic triangle, because its repetition ultimately fills the entire surface of the sphere. It comes, of course, in right- and left-handed forms. If the vertices of the characteristic triangle are named P, Q, R, respectively, the triangle may be named $(p\ q\ r)$, a designation relating it to the measures of its angles (see Fig. 16). Using radian measure for the angles simplifies matters. In Fig. 16, the angle at R is 90°, hence $R = \pi/2$ radians; the angle at P is π/p radians, and the angle at Q is π/q radians.

For the regular spherical models the values of p, q are 3,4,5. Thus for the tetrahedral case, $p = q = 3$. You should know that $\pi/3$ radians = 60°. For the octahedral case, $p = 3$ and $q = 4$ and $\pi/4$ radians = 45°. For the icosahedral case $p = 3$ and $q = 5$ and $\pi/5$ radians = 36°.

Some understanding of the geometry of the spherical triangle is useful here. Both

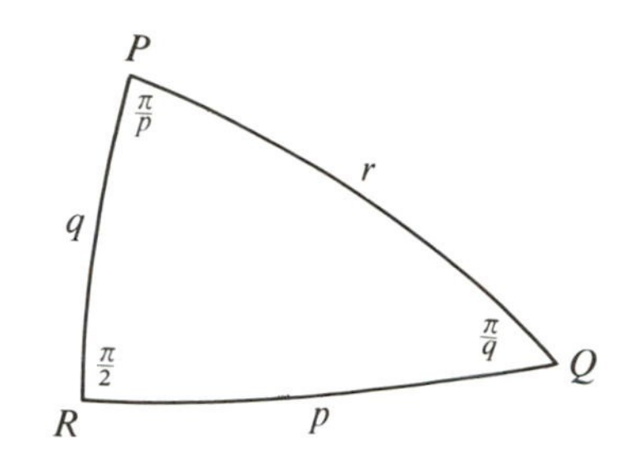

Fig. 16. A spherical triangle.

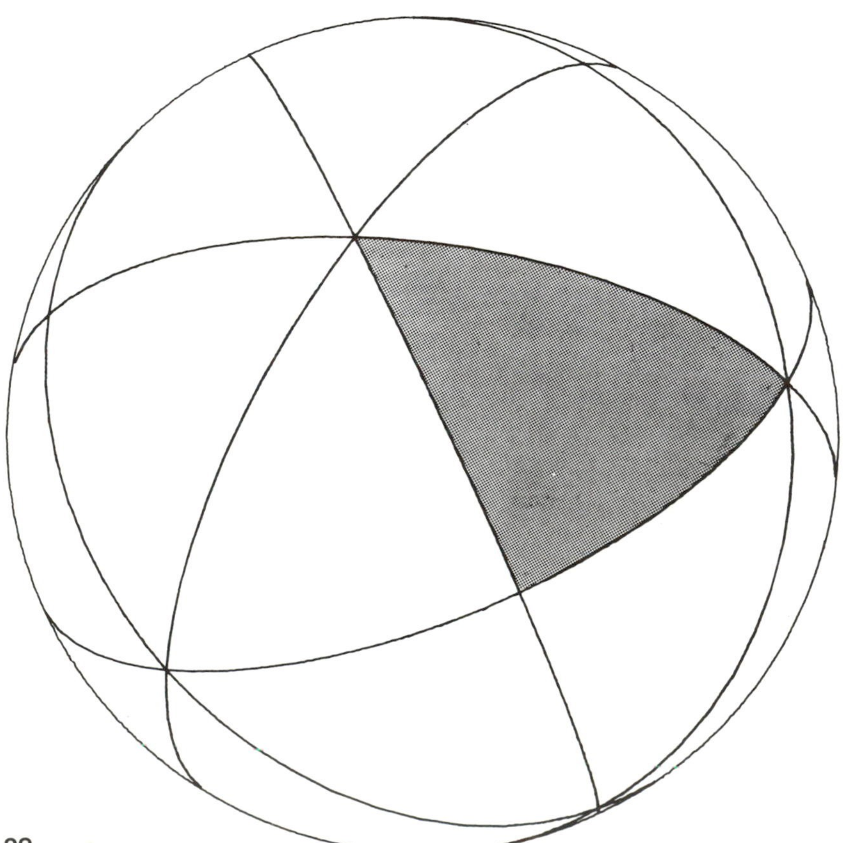

Plate 9. (2 3 3).

Plate 10. (2 3 4).

Plate 11. (2 3 5).

the sides and the angles of a spherical triangle are measured by angular measure, either in degrees or radians. The sides of a spherical triangle are arcs of three great circles, each with an angular measure that is the same as the angle each arc subtends at the center of the sphere. These sides are named with small letters: p, q, r or a, b, c. The angles of a spherical triangle are not arcs of great circles but are really the dihedral angles between the two planes determined by the center of the sphere and two great circle arcs forming sides of the spherical triangle. These angles are named with capital letters P, Q, R or A, B, C or, as in Fig. 1, Q_0, Q_1, Q_2.

Look now at Fig. 17a which shows the three kinds of characteristic triangles found in the regular spherical models. The numerals in Fig. 17a designate the points whose reflections generate the polyhedrons designated by the same numerals in the list given with Fig. 17a. It will be seen that other points suitably chosen on the sides of these characteristic triangles or within them will generate all the remaining semiregular polyhedrons.

What is meant by "suitably chosen"? One choice to make is that of the bisector of the angle π/p or π/q. In the tetrahedral case, if $\pi/3$ is bisected, the bisecting arc meets the opposite side in a point that, along with its images, generates the vertices of the truncated tetrahedron. Because two angles each are $\pi/3$, this can be done in two ways, but they both generate the same polyhedron. In the octahedral case, if $\pi/3$ is bisected, the bisecting arc meets the opposite side in a point that, along with its image, marks the vertices of the truncated octahedron. The bisector of the angle $\pi/4$ leads to the truncated cube. In the icosahedral case, the bisector of the angle $\pi/3$ leads to the truncated icosahedron, and the bisector of the angle $\pi/5$ leads to the truncated dodecahedron. Thus all five truncated forms of the regular polyhedrons are accounted for (see Fig. 17b). These may now be studied in conjunction with Fig. 17a and Plates 12–16. If you study these by looking at the three regular spherical models, you will notice that for all of these you must use your imagination.

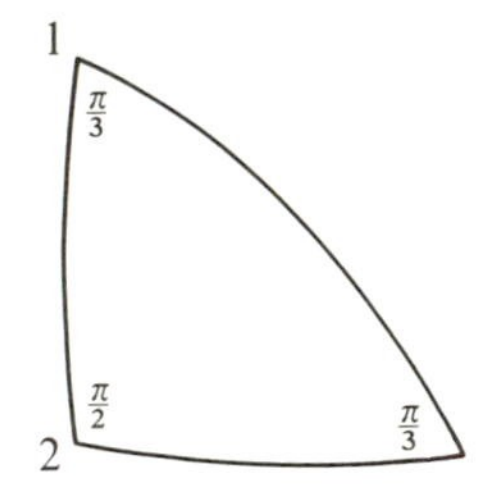

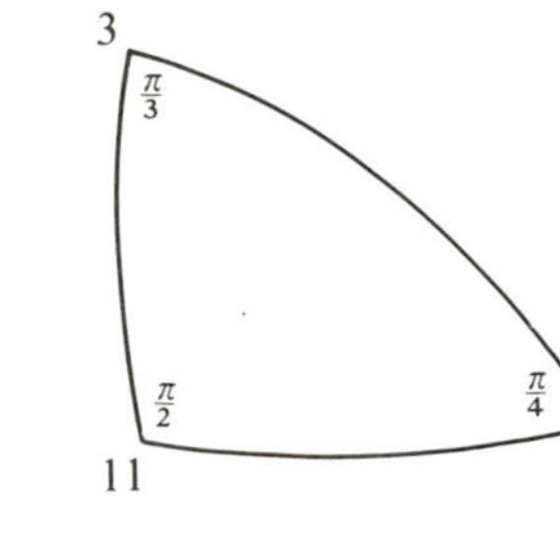

1 Tetrahedron
2 Octahedron
3 Cube
4 Icosahedron
5 Dodecahedron
11 Cuboctahedron
12 Icosidodecahedron

Fig. 17a. The three characteristic triangles with vertex points identified in relation to spherical polyhedrons.

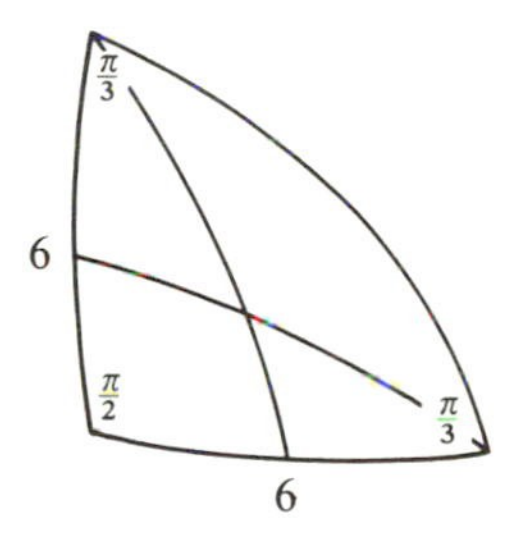

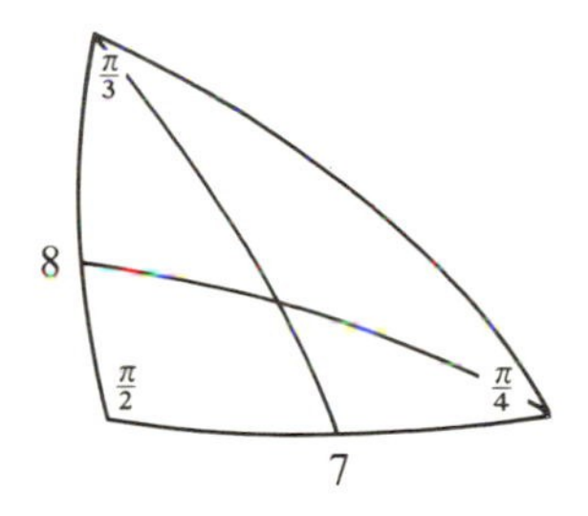

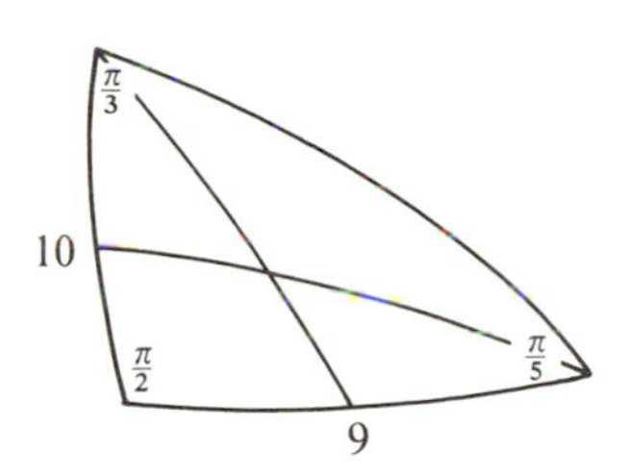

6 Truncated tetrahedron
7 Truncated octahedron
8 Truncated cube
9 Truncated icosahedron
10 Truncated dodecahedron

Fig. 17b. The three characteristic triangles with acute angles bisected showing the relationship to five semiregular polyhedrons, the truncated regular solids.

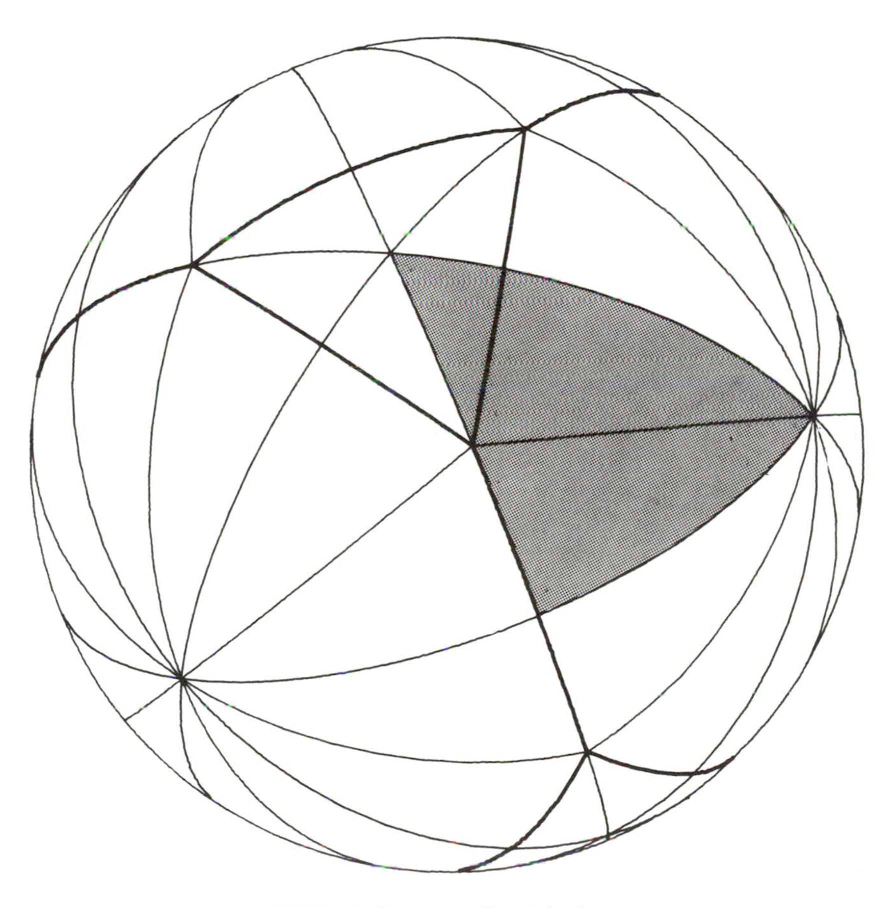

Plate 12. Truncated tetrahedron.

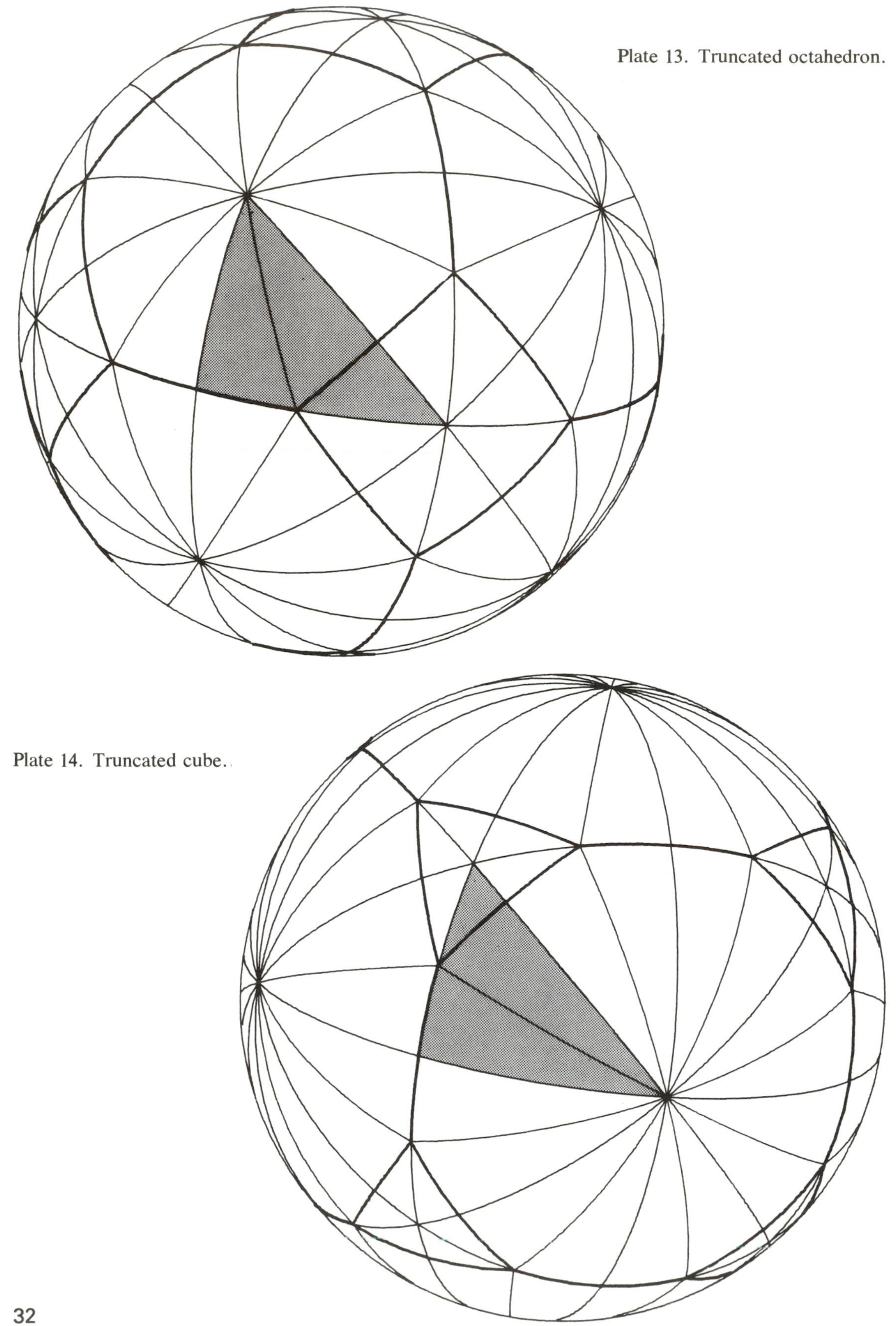

Plate 13. Truncated octahedron.

Plate 14. Truncated cube.

Plate 15. Truncated icosahedron.

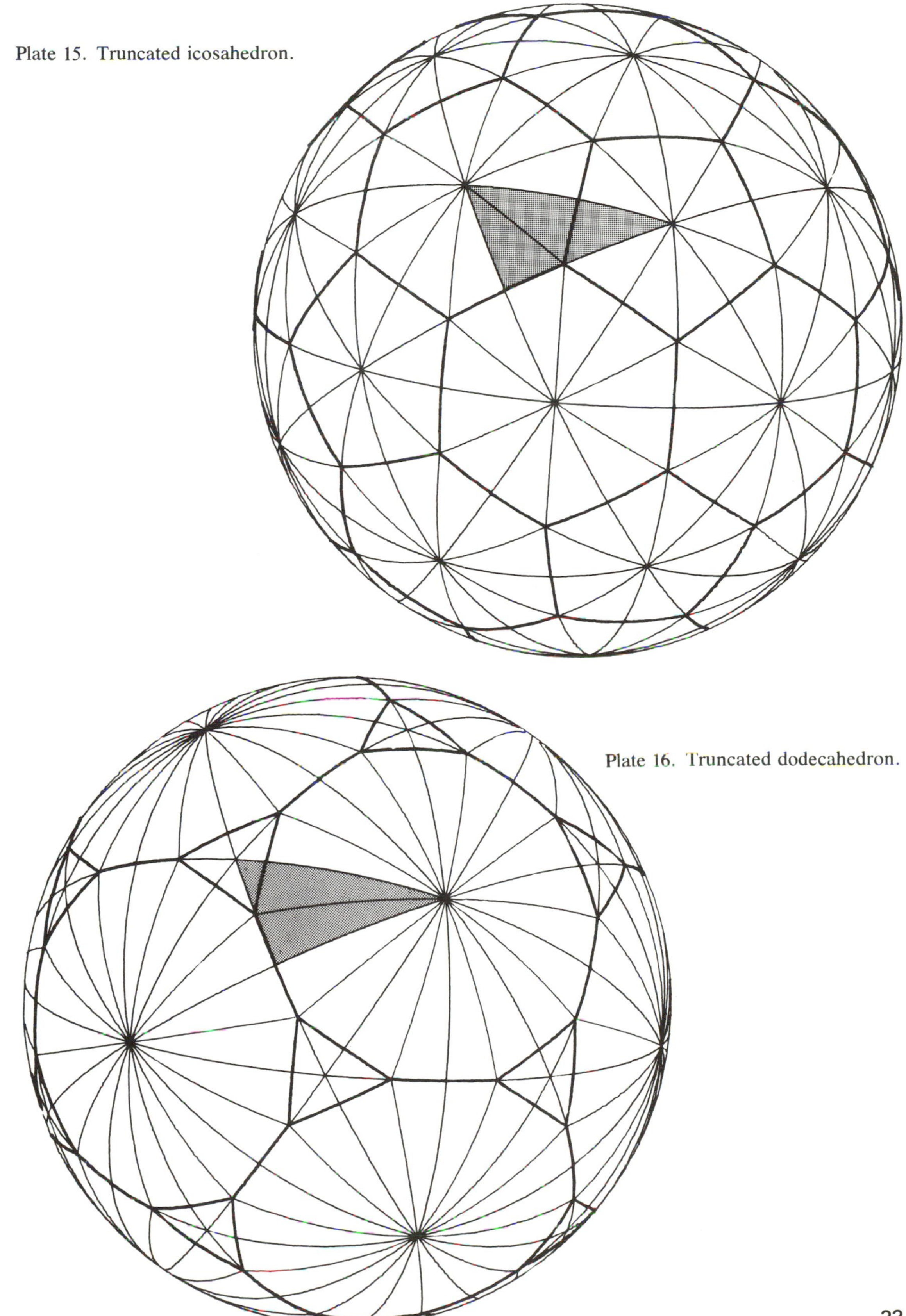

Plate 16. Truncated dodecahedron.

The five truncated regular spherical models

Instead of relying solely on your imagination to see these results, it is far more satisfying to make the spherical models of the five truncated regular polyhedrons. Each has its own required number of triangles, squares, and pentagons as faces, all of which have been seen before, but now hexagons, octagons, and decagons also appear. You will need the values of a and r for these. As before, e will in each case remain 2 units of length, making $\frac{1}{2}e = 1$. The values of a and r were given in Table 2 at the end of Section I. You are invited to do the calculations. The numerals designating each polyhedron correspond to those used in Fig. 17b.

The layouts for 6, 7, 8, 9, 10 are given in Figs. 18–22. These now need no further comment. You have presumably already made 11 and 12. The bands are designed in the usual way with any suitable radius.

As the number of bands needed for these models increases the work takes longer. But the process is still the same. Make spherical polygons first, of the types needed, and cement them together following the facial arrangements appropriate to each. You realize, of course, that the spherical polygons of one model are not interchangeable with those of another, because each model has its own radius. This changes its value from one model to the next. Photos 6–10 show complete spherical models. If you are not inclined to make all of them, do at least a partial model. This will aid you to understand the relationships. Even for a partial model the parts are remarkably rigid from the very start because each spherical right triangle is individually rigid.

The layouts, as you can see, have been simplified by omitting some of the geometrical constructional details. Calculated data may be used instead; these are given in Table 3 at the end of this section.

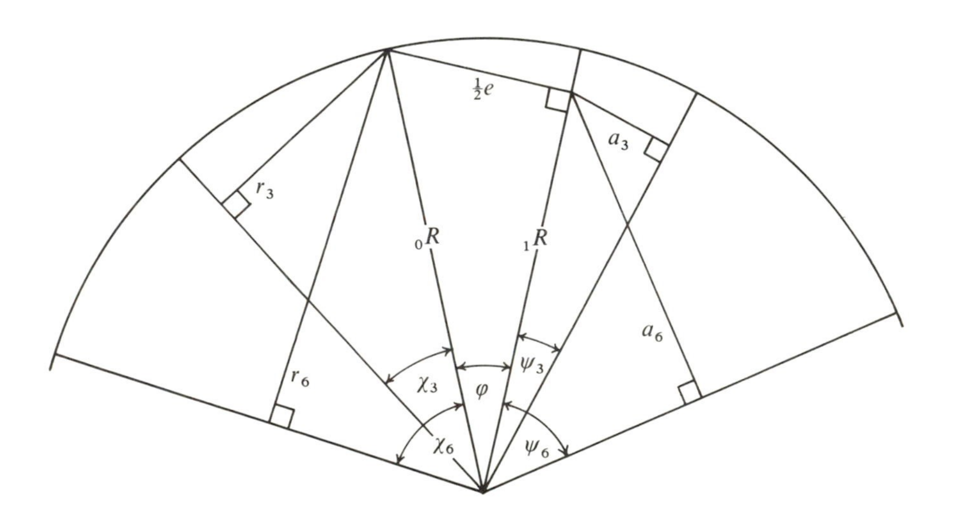

Fig. 18. Layout for truncated tetrahedron.

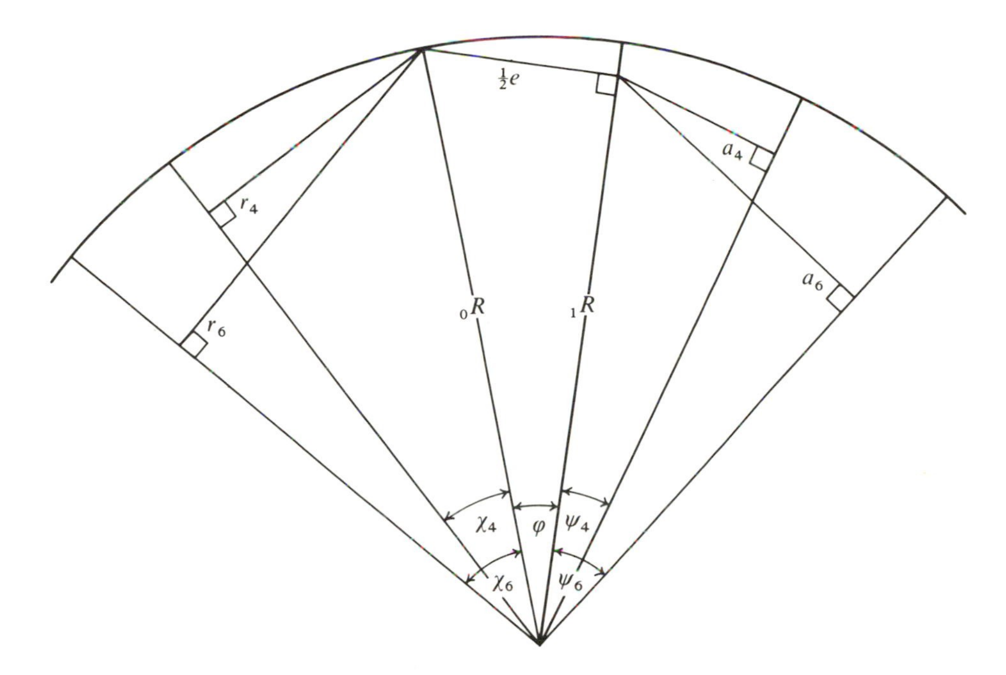

Fig. 19. Layout for truncated octahedron.

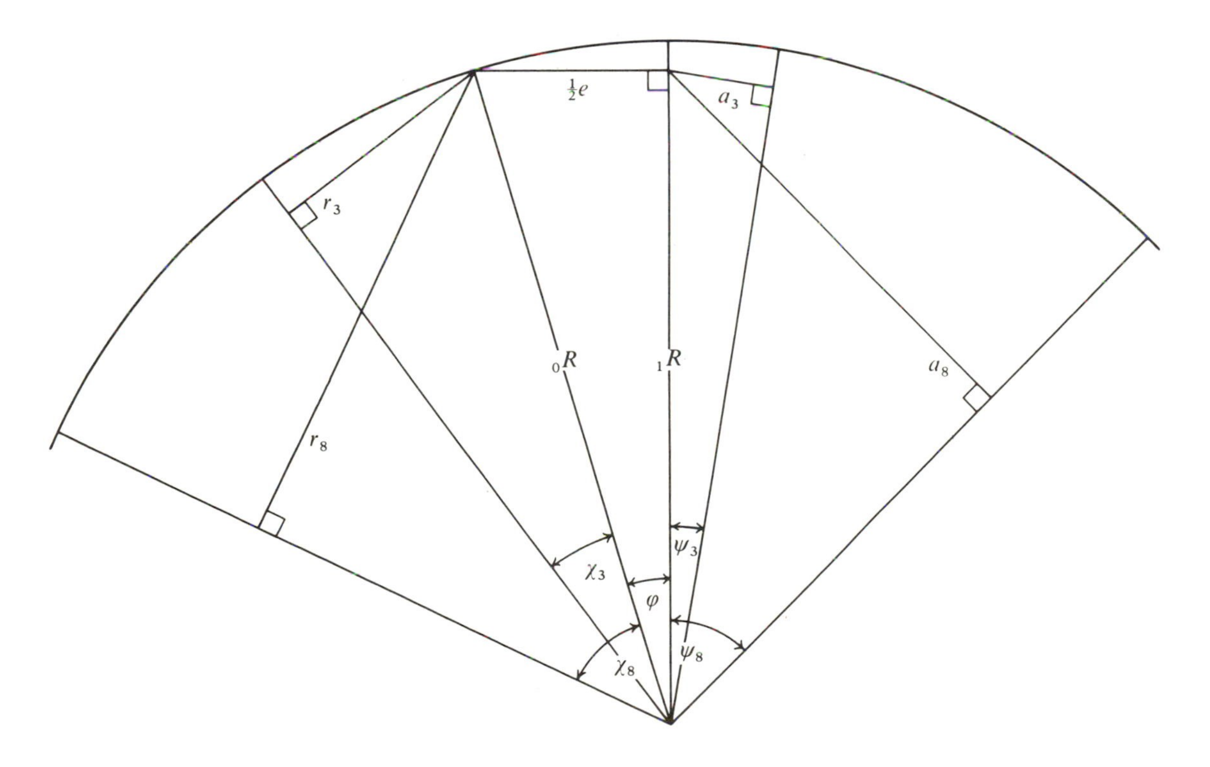

Fig. 20. Layout for truncated cube.

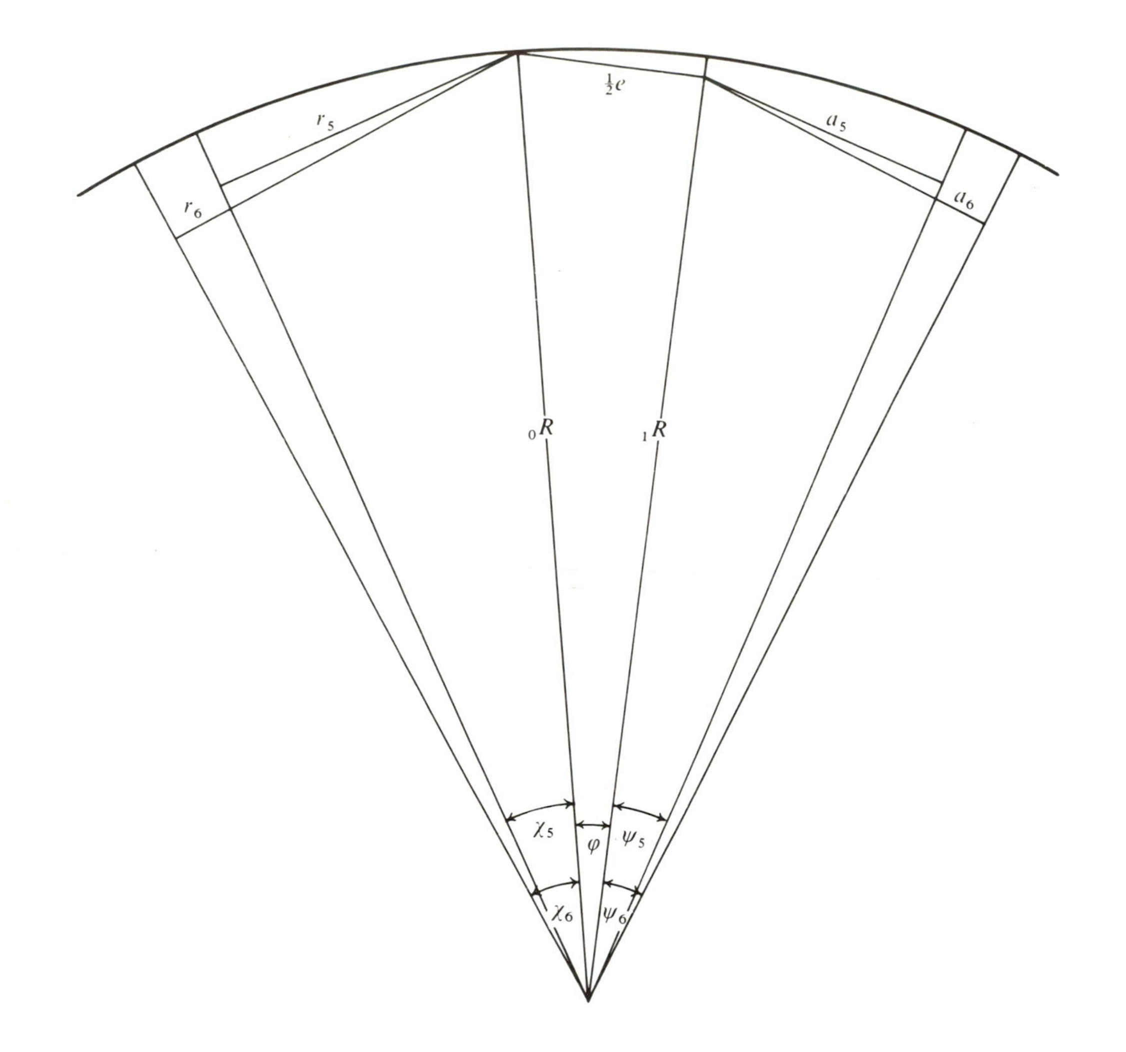

Fig. 21. Layout for truncated icosahedron.

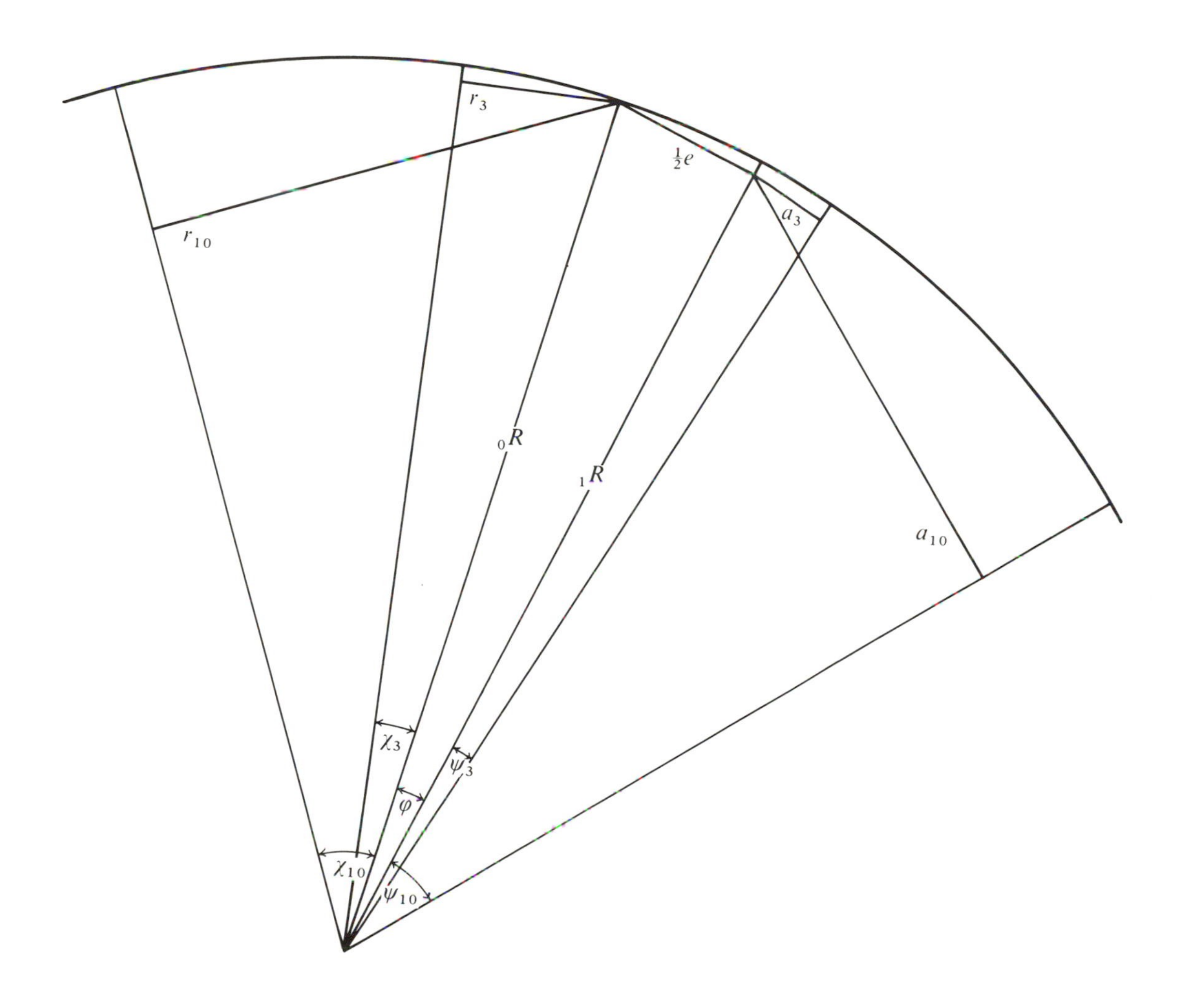

Fig. 22. Layout for truncated dodecahedron.

Photo 6. Truncated tetrahedron.

Photo 7. Truncated octahedron.

Photo 8. Truncated cube.

Photo 9. Truncated icosahedron.

Photo 10. Truncated dodecahedron.

The rhombic spherical models

Looking back at Fig. 17b, you see that the angle $\pi/2$ was not bisected. Would the point derived from this bisector be a suitable point for some of the remaining semiregular solids? The answer is yes, and the results of this choice are shown in Fig. 23. In the tetrahedral case the cuboctahedron reappears; in the other two cases the rhombicuboctahedron and the rhombicosidodecahedron are generated. These are shown in Plates 17–19. They bear the numerals 11, 13, 14 in the list given with Fig. 23. You can examine the three regular spherical models to see how these relationships are present. However, you may find it more satisfying to make the models.

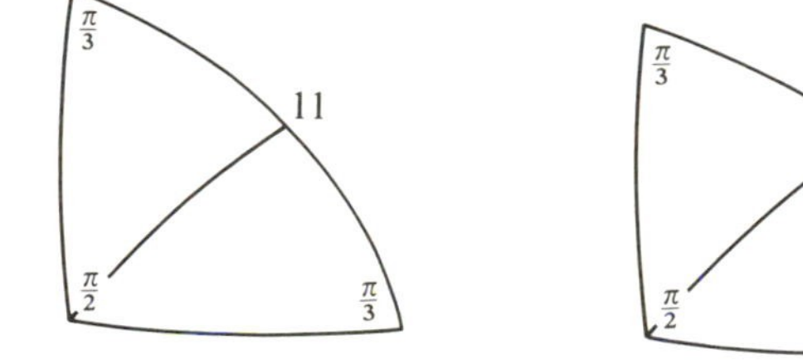

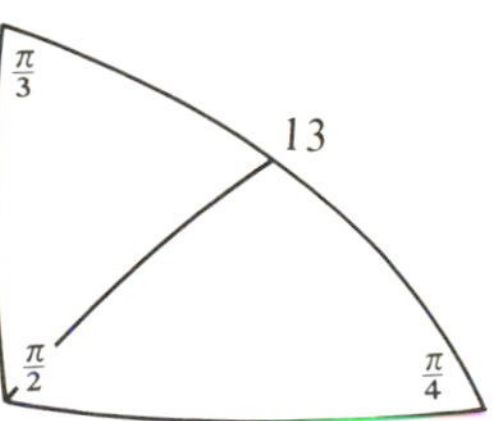

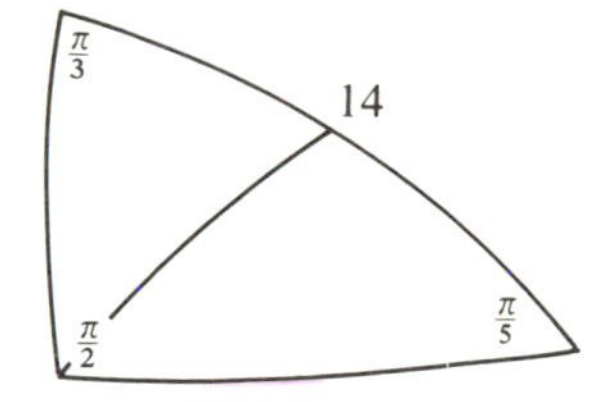

11 Cuboctahedron
13 Rhombicuboctahedron
14 Rhombicosidodecahedron

Fig. 23. The three characteristic triangles with the right angle bisected showing the relationship to three semiregular polyhedrons.

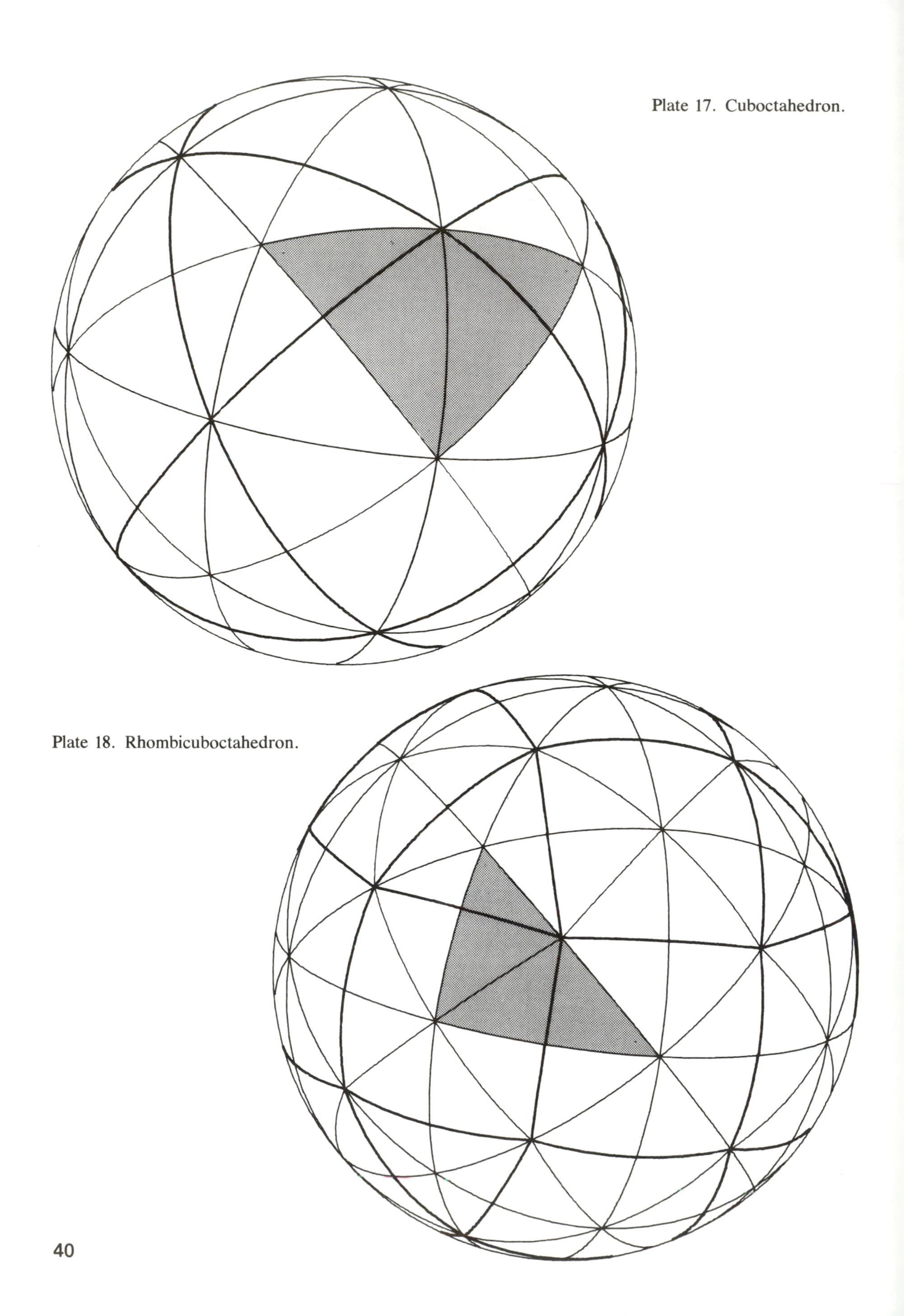

Plate 17. Cuboctahedron.

Plate 18. Rhombicuboctahedron.

Photo 11. Cuboctahedron.

Photo 12. Icosidodecahedron.

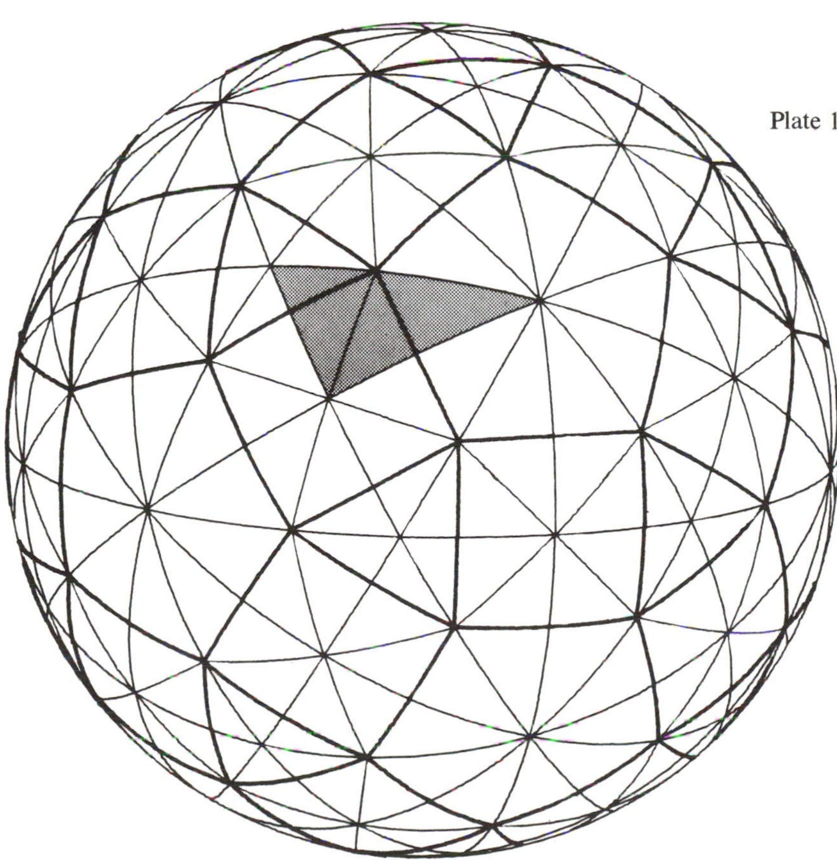

Plate 19. Rhombicosidodecahedron.

Photo 13. Rhombicuboctahedron.

Photo 14. Rhombicosidodecahedron.

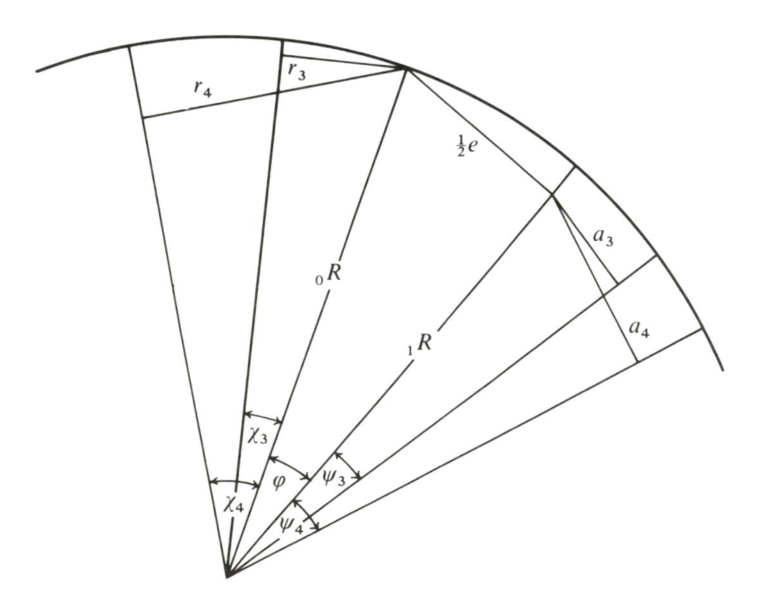

Fig. 24. Layout for rhombicuboctahedron.

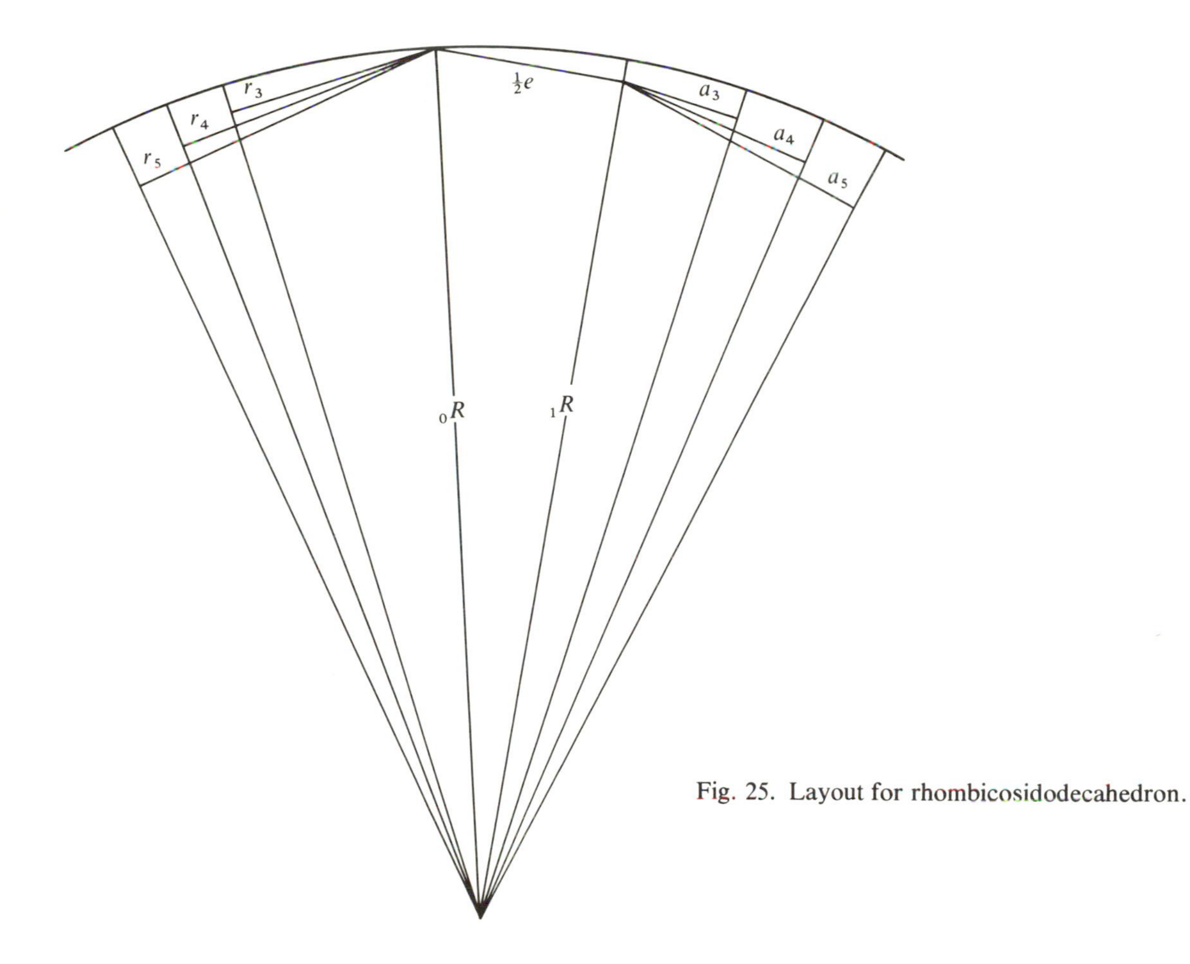

Fig. 25. Layout for rhombicosidodecahedron.

Figures 24 and 25 show the layouts from which the bands are derived for making 13 and 14. In 13 the rhombic squares are only significant so far as the symmetry is concerned. All the squares in this model have the same linear and spherical measures. In 14, three types of faces appear calling for three kinds of bands. This fact undoubtedly adds a great deal to its attractiveness.

This now leaves 15 and 16 to be investigated, the rhombitruncated cuboctahedron and the rhombitruncated icosidodecahedron. Also 17 and 18, the snub cube and the snub dodecahedron, are needed to complete the list of thirteen semiregular polyhedrons.

The rhombitruncated spherical models

Since all three angle bisectors have now been used, you might suspect that the point at which all three meet on the interior of a characteristic triangle, namely the incenter, should generate the rhombitruncated forms. This is indeed what happens. In the tetrahedral case the truncated octahedron turns up again; in the other two, the rhombitruncated forms appear (see Plates 20–22 and Fig. 26).

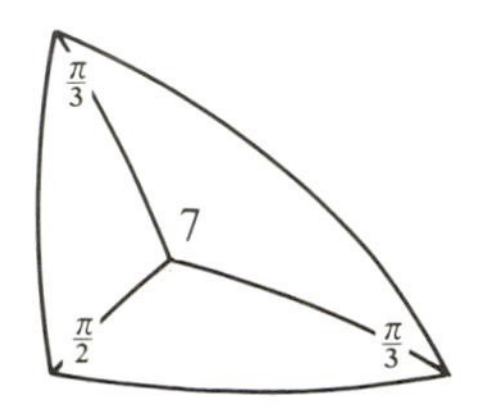

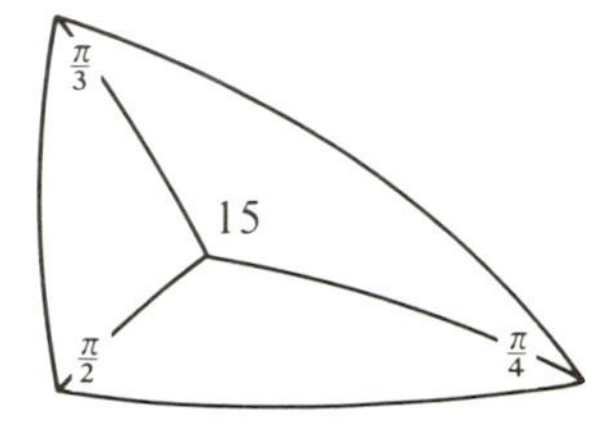

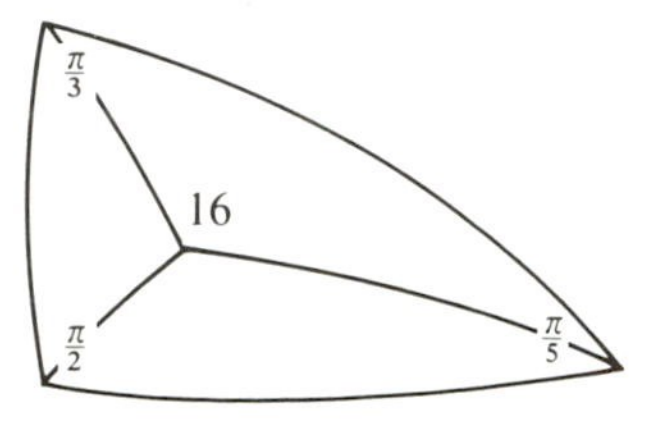

7 Truncated octahedron
15 Rhombitruncated cuboctahedron
16 Rhombitruncated icosidodecahedron

Fig. 26. The three characteristic triangles with their incenter point showing the relationship to three more semiregular polyhedrons.

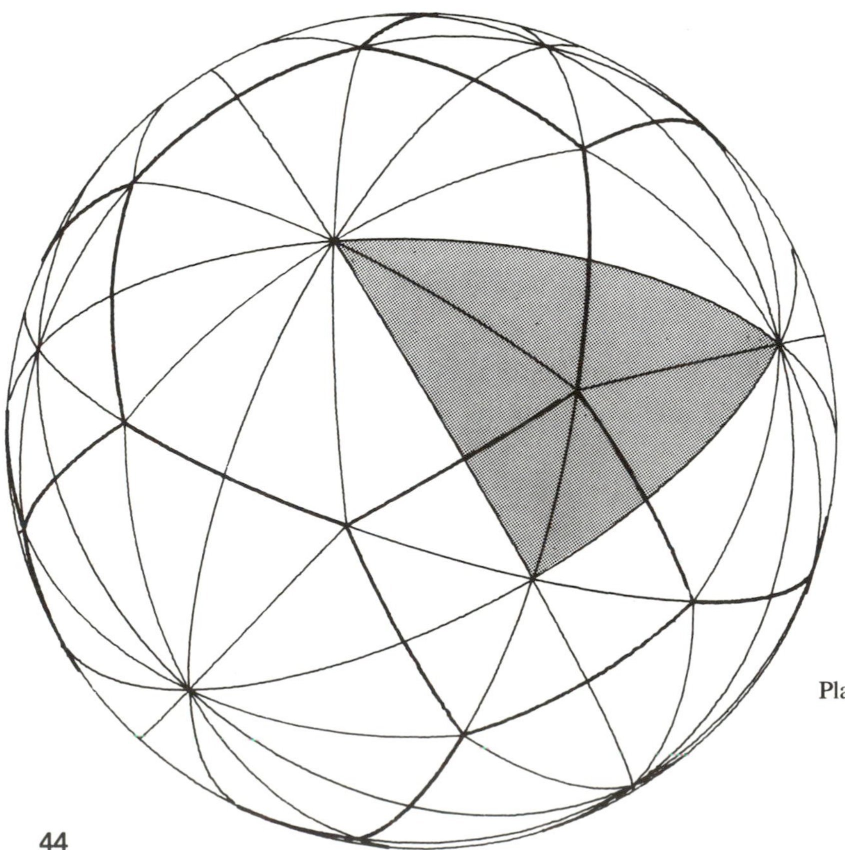

Plate 20. Truncated octahedron.

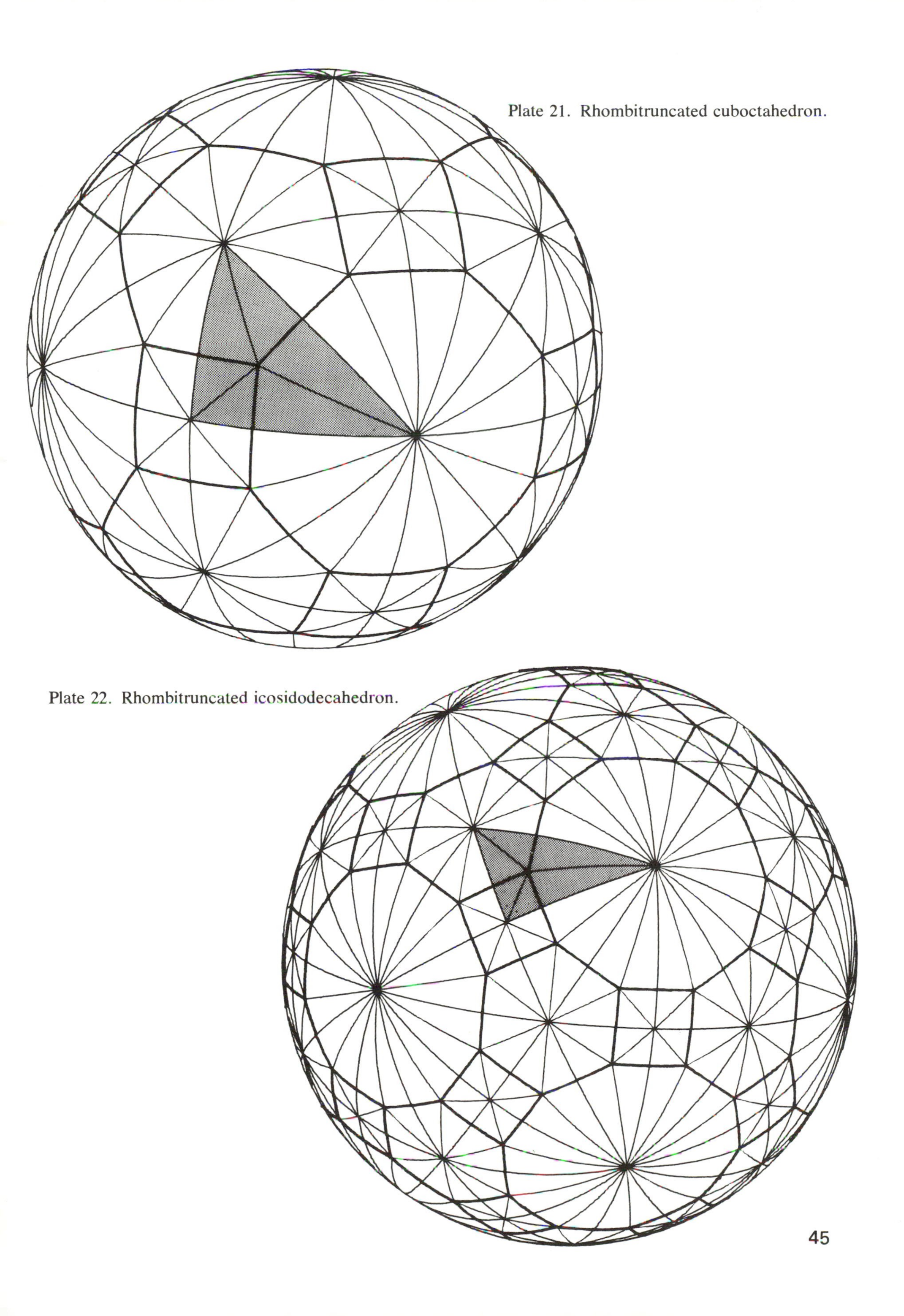

Plate 21. Rhombitruncated cuboctahedron.

Plate 22. Rhombitruncated icosidodecahedron.

Photo 15. Rhombitruncated cuboctahedron.

Photo 16. Rhombitruncated icosidodecahedron.

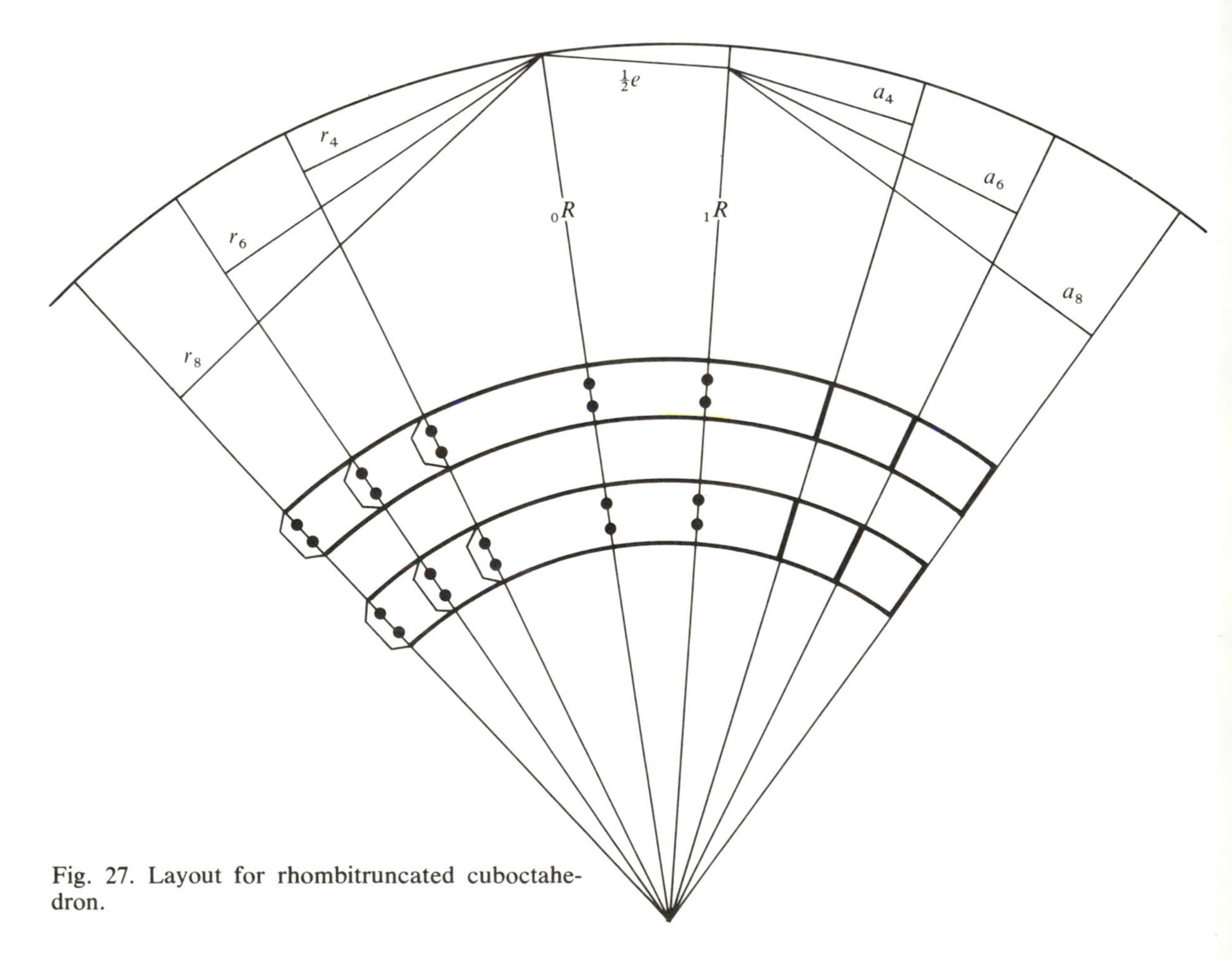

Fig. 27. Layout for rhombitruncated cuboctahedron.

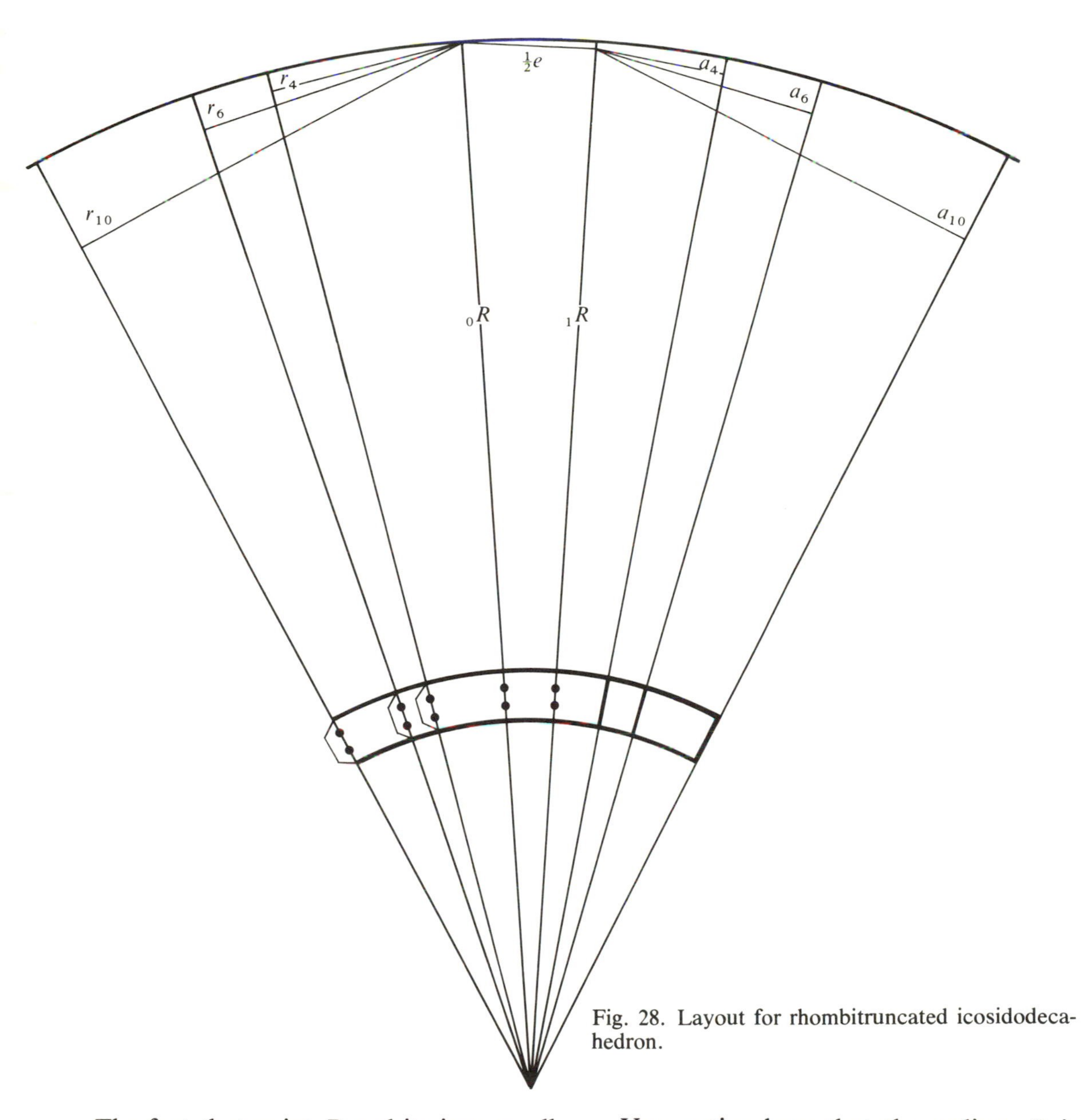

Fig. 28. Layout for rhombitruncated icosidodecahedron.

The fact that point P and its images all lie on the interior of a characteristic triangle makes it harder for the imagination to see the faces of the respective polyhedrons on the regular spherical models, so again it is more satisfying to make models that you can hold in your hands and turn this way and that to see the results with your eyes. The layouts are shown in Figs. 27 and 28.

You notice here that the radius $_0R$ is becoming relatively large. As already stated the bands may be made smaller: A radius of 3 in. or about 7.5 cm is quite suitable. Long bands have a tendency to become bent or bowed in a paper model. The bands in Figs. 27 and 28 are shown one overlaid on the others. The templates must of course be made separately.

The snub forms as spherical models

The snubs 17 and 18 are each generated from a specially chosen point within the characteristic triangles needed for each. Its exact position or location is rather a complicated affair. If you are interested in seeing how to locate this point on the flat faces of a cube and a dodecahedron, respectively, you may consult L. Lines, *Solid geometry,* pp. 175–84. For spherical models this point would then have to be projected onto the surface of a circumscribing sphere. Plate 23 shows the snub cube. Plate 24 illustrates the fact that the point to be reflected can be chosen inside the mirror image of the characteristic triangle, thus leading to the mirror-image polyhedron.

The spherical models themselves follow easily from their respective layouts (see Figs. 29 and 30). All the triangles of the snub cube are the same, so these are assembled and cemented to the spherical squares by following either a right- or left-handed symmetry. The same is true of the snub dodecahedron relative to its spherical pentagons. Photos 17 and 18 show the complete models. The snub dodecahedron is especially worth making, because it will be featured in another role in Section IV of this book.

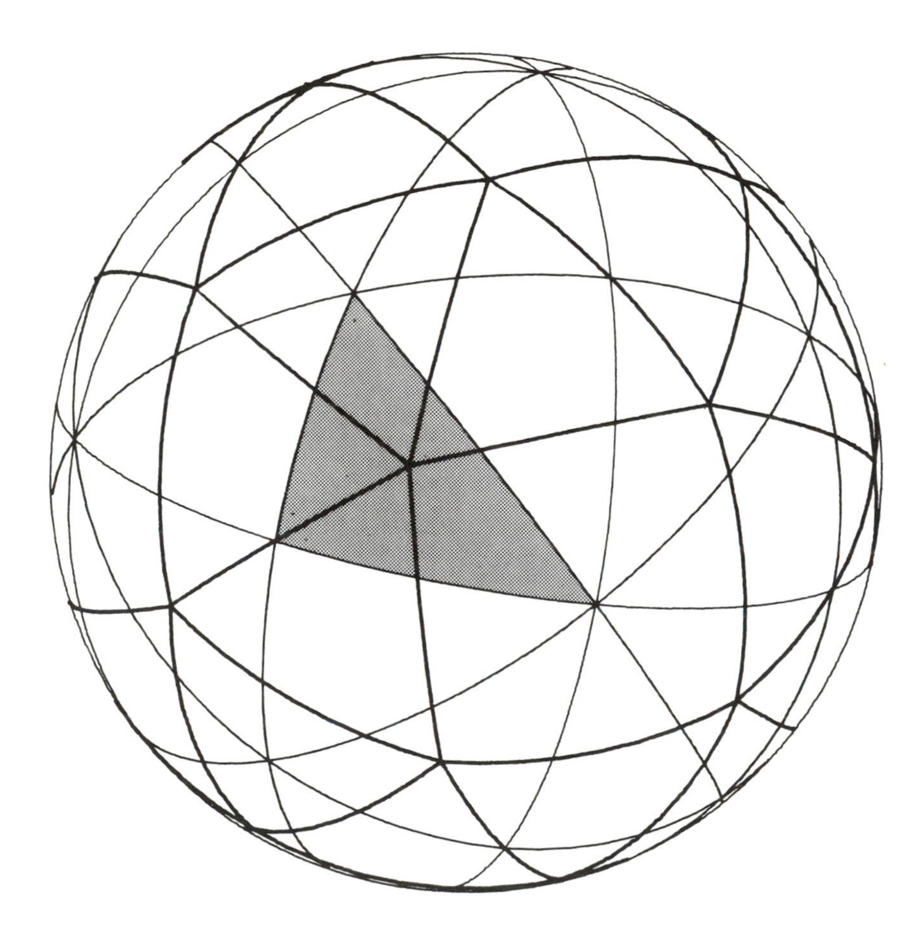

Plate 23. Snub cube.

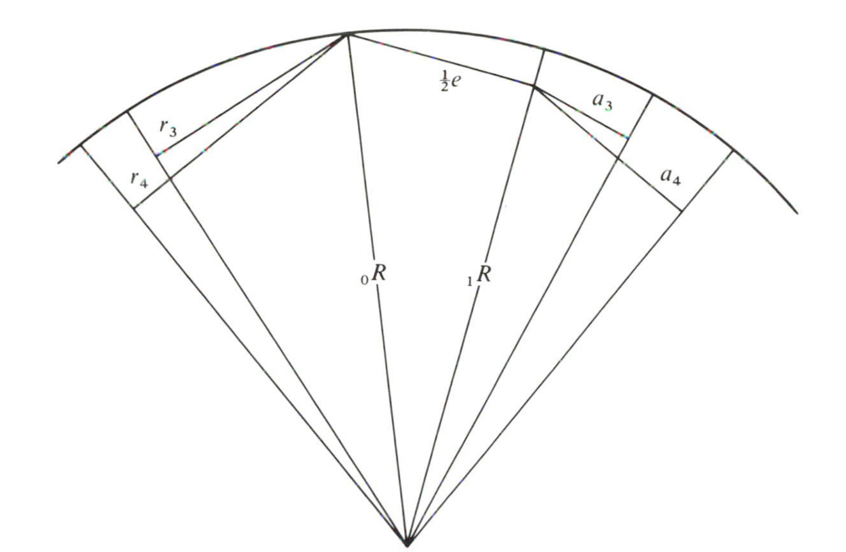

Fig. 29. Layout for snub cube.

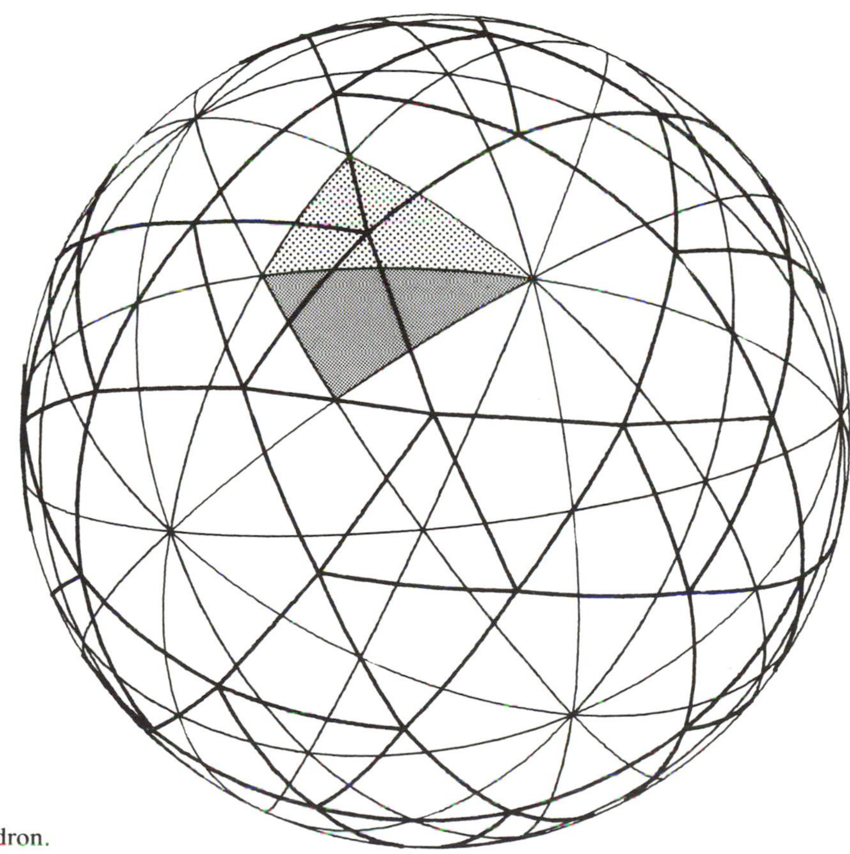

Plate 24. Snub dodecahedron.

Photo 17. Snub cube.

Photo 18. Snub dodecahedron.

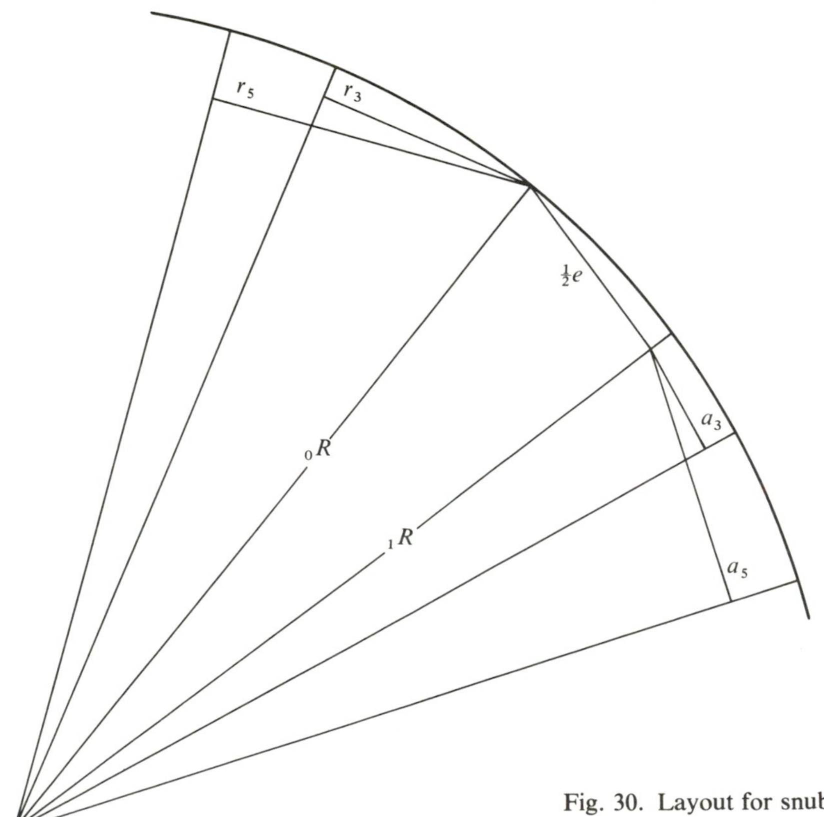

Fig. 30. Layout for snub dodecahedron.

The spherical duals

You may by now have noticed that the duals of all these polyhedrons are present in the models already constructed, but they are present here as projected onto the same circumscribing sphere. Plates 25–27 show heavy lines marking the edges of the triakistetrahedron, the rhombic dodecahedron, and the rhombic triacontahedron. These are, respectively, the duals of the truncated tetrahedron, the cuboctahedron, and the icosidodecahedron. The heavy lines are only illustrative of course. If you pick up your model of the regular spherical octahedron, you will see that you can fix your attention upon a group of characteristic triangles that composes an equilateral triangle or upon another group that composes a square. The first group has six characteristic triangles, the other has eight. However, you can also fix your attention upon a group of four characteristic triangles that composes an equilateral rhomb. This rhomb appears a total of 12 times to cover the surface of the sphere. Notice that the total enumeration of characteristic triangles comes out the same in all three instances: $6 \times 8 = 48$; $8 \times 6 = 48$; $4 \times 12 = 48$. On the regular spherical icosahedron the groupings become triangles, pentagons, or rhombs. The enumeration comes out as: $6 \times 20 = 120$; $10 \times 12 = 120$; $4 \times 30 = 120$.

If you examine the spherical cuboctahedron, the rhomb appears as being composed of eight right spherical triangles

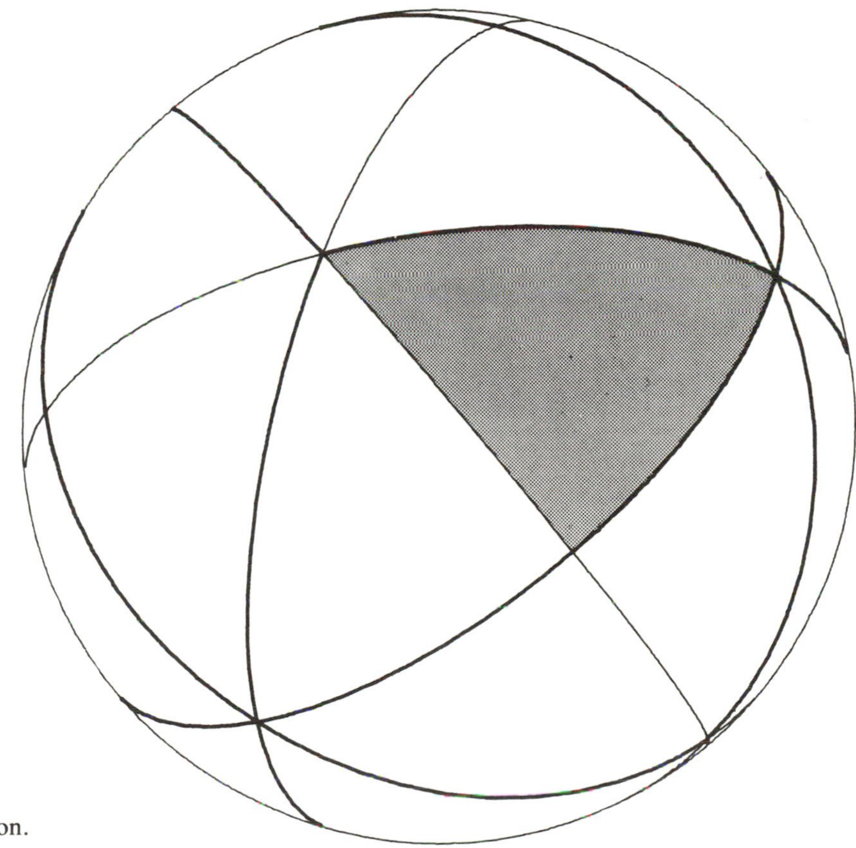

Plate 25. Triakistetrahedron.

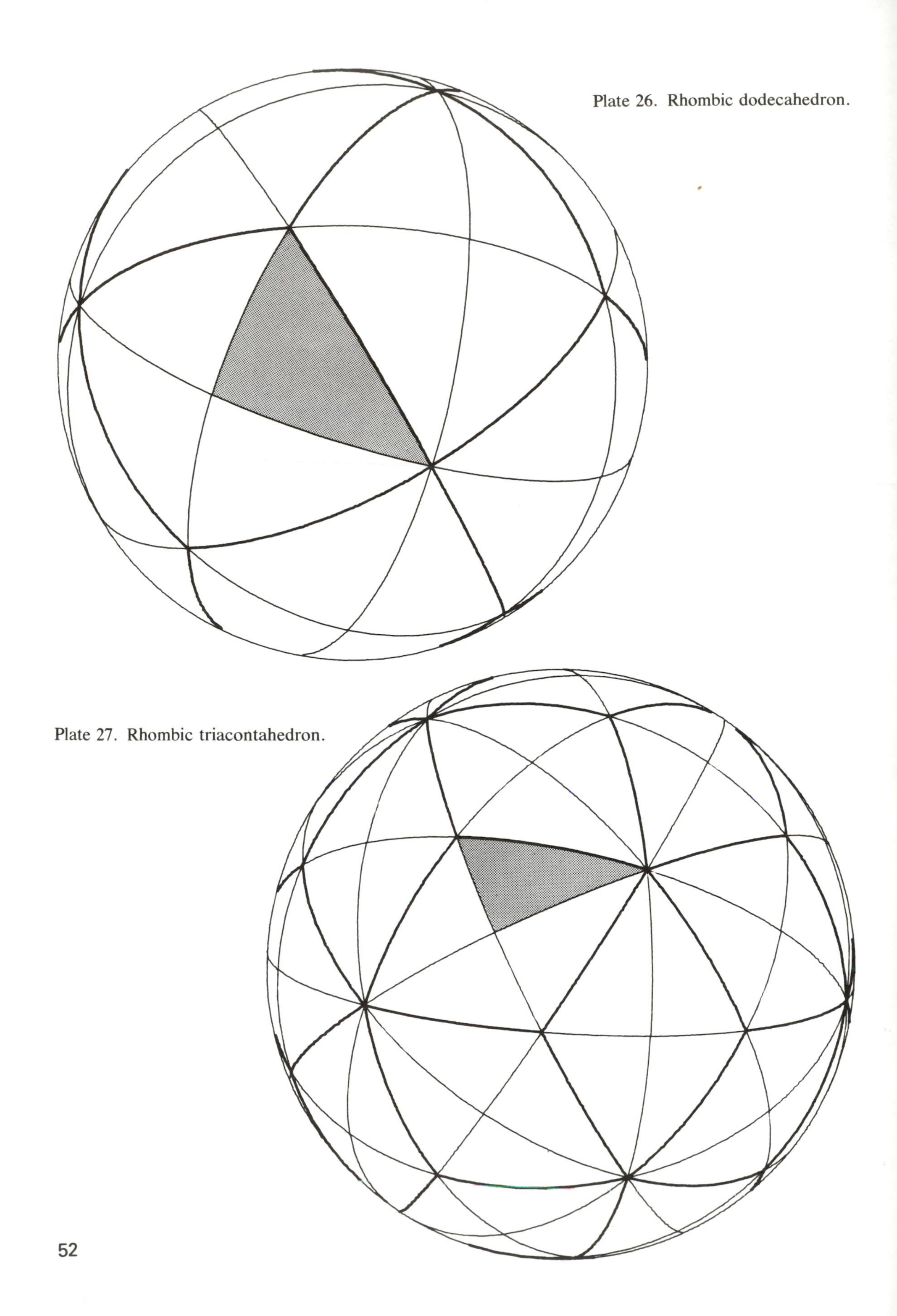

Plate 26. Rhombic dodecahedron.

Plate 27. Rhombic triacontahedron.

Table 3. *The thirteen semiregular solids*[a]

Polyhedron (for $e = 2$)		${}_0R$	${}_1R$	${}_2R$		χ	ϕ	ψ
6 Truncated tetrahedron	t{3, 3}	2.345	2.121	2.041 1.225	{3} {6}	29.496 58.518	25.239	15.793 54.736
7 Truncated octahedron	t{3, 4}	3.162	3.000	2.828 2.449	{4} {6}	26.565 39.232	18.435	19.471 35.264
8 Truncated hexahedron	t{4, 3}	3.558	3.414	3.365 2.414	{3} {8}	18.939 47.266	16.325	9.736 45.000
9 Truncated icosahedron	t{3, 5}	4.956	4.854	4.655 4.535	{5} {6}	20.077 23.800	11.641	16.472 20.905
10 Truncated dodecahedron	t{5, 3}	5.939	5.855	5.826 4.980	{3} {10}	11.211 33.017	9.694	5.660 31.717
11 Cuboctahedron	{3} {4}	2.000	1.732	1.633 1.414	{3} {4}	35.264 45.000	30.000	19.471 35.264
12 Icosidodecahedron	{3} {5}	3.236	3.078	3.023 2.753	{3} {5}	20.905 31.717	18.000	10.812 26.565
13 Rhombicuboctahedron	r{3} {4}	2.798	2.613	2.549 2.414	{3} {4}	24.374 30.361	20.941	12.764 22.500
14 Rhombicosidodecahedron	r{3} {5}	4.466	4.353	4.314 4.236 4.129	{3} {4} {5}	14.985 18.462 22.393	12.939	7.623 13.283 18.435
15 Rhombitruncated cuboctahedron	t{3} {4}	4.635	4.526	4.414 4.182 3.828	{4} {6} {8}	17.764 25.561 34.316	12.459	12.764 22.500 32.236
16 Rhombitruncated icosidodecahedron	t{3} {5}	7.604	7.538	7.472 7.337 6.881	{4} {6} {10}	10.718 15.249 25.186	7.557	7.623 13.283 24.097
17 Snub cube	s{3} {4}	2.687	2.494	2.427 2.285	{3} {4}	25.446 31.751	21.845	13.383 23.634
18 Snub dodecahedron	s{3} {5}	4.312	4.194	4.154 3.962	{3} {5}	15.534 23.240	13.411	7.912 19.158

[a]The linear measures ${}_0R$, ${}_1R$, ${}_2R$ are in the same units of length as the edge length $e = 2$; the angular measures χ, ϕ, ψ, are in degrees.

of two different kinds. Here $(6 \times 8) + (8 \times 6) = (8 \times 12)$. The situation for the spherical icosidodecahedron is $(6 \times 20) + (10 \times 12) = (8 \times 30)$.

It will be left for your inspection to see how the other semiregular duals are to be found in the other models. You should be able to pick out the number and the shape of each face if you remember that each dual has only one kind of face. See Holden, pp. 54–5, or Pugh, *Polyhedra, a visual approach,* pp. 43–4, for a complete list of duals.

Summary

This section on the semiregular models can now be summarized by the data given in Table 3. You are invited to verify all the results using the mathematical relationships given at the end of Section I.

III. Variations

All through Sections I and II of this book the circular bands of paper used for making the regular and semiregular spherical models were all right spherical triangles. Once the arc lengths were determined, they automatically took on the correct shapes to fix the dihedral angles between the bands and, being repeated a sufficient number of times, they covered the entire surface of the sphere. This is what is meant by spherical tessellation.

The question might now be asked: Can triangles other than right spherical triangles be used for spherical tessellation? The answer is yes, and no doubt you may have already seen the possibility of shortening both the amount of time and work involved in making a model by simply combining a pair of right triangles, a right-handed and a left-handed one, to make one isosceles triangle. This means dropping a but keeping r and using e instead of $\frac{1}{2}e$. This changes the central angles to 2ϕ and χ, eliminating ψ. A suitable arrangement is to use 2ϕ with χ on either side of it when designing the bands needed for models of this kind. They may be called variations of the spherical models seen so far.

Regular and semiregular variations

For the regular polyhedrons the number of bands is reduced by half. For the tetrahedron only twelve bands are needed. The cube and the dodecahedron are here found to be distinctly different models from the octahedron and the icosahedron. For the cube and the octahedron, twenty-four bands are needed, whereas for the dodecahedron and the icosahedron sixty are needed. The dodecahedron is especially worth making in this way, because it is the first model that begins to take on the appearance of a geodesic dome. More will be said about this in Section IV.

The treatment given to the regular spherical models can be applied to all the semiregular ones. The polygons take on the appearances shown in Fig. 31.

For a triangle face, even r can be eliminated and the result is an equilateral spherical triangle with each arc 2ϕ, which can be used in combination with the other polygon faces. You realize of course that the value of 2ϕ changes from one model to the next.

Photos 19 and 20 show only a sampling of models done in this way. They are meant to stimulate your own ideas, to stir up some of your own creative spirit, to leave the way open for your own experimentation.

If in the regular and semiregular polyhedra, both a and r are eliminated, this leaves e alone with 2ϕ as the central angle. The bands will then have the same number of arcs as the polygon has sides, each with a central angle of 2ϕ. You will find, however, that this sacrifices some rigidity, since only the triangle itself remains rigid. The tetrahedron done in this way needs only four bands. The model is not very attractive nor is it very satisfactory because the arcs are comparatively long. Other polyhedra are a bit more satisfactory in particular those that have triangles for faces, because these add rigidity to the overall model. When done on a small scale they can be both attractive and satisfactory. They make useful decorative devices. In Section V further uses will be suggested when miscellaneous models derived from spherical ones will be treated.

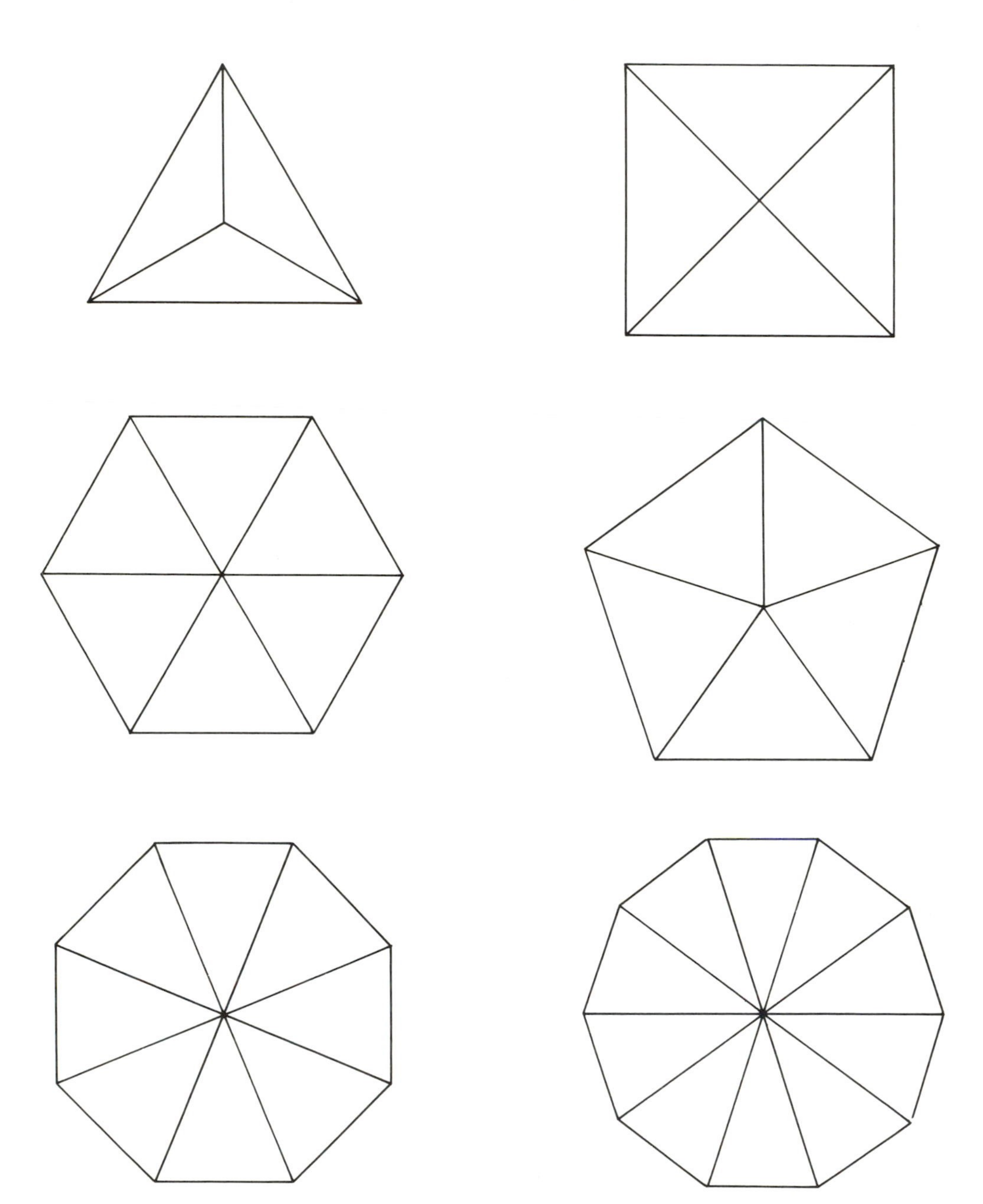

Fig. 31. The six regular polygons omitting a but retaining r.

Photo 19. *Clockwise from the top*: icosahedron, cuboctahedron, octahedron (omitting a and r), tetrahedron, and octahedron.

Photo 20. *Clockwise from the top*: icosahedron, cube, icosidodecahedron (variation), triakistetrahedron, and tetrahedron.

Star-faced spherical models

The next question you might raise is: Can spherical models be made for star-faced polyhedrons? Many nonconvex uniform polyhedrons have pentagrams, octagrams, and decagrams as faces, but the hexagram strangely enough does not appear. You may be delighted now to know that all of these stars can be drawn on the plane faces of the regular and semiregular polyhedrons and then projected by central or gnomonic projection onto the surface of a circumscribing sphere. This can be done using ordinary geometry and trigonometry. There is no real need for the use of spherical trigonometry.

Dodecahedron with pentagrams

A particularly attractive model to begin with is the regular dodecahedron with a pentagram drawn on each pentagon face. Figure 32 shows one face of the dodecahedron with its inscribed pentagram. The small letters *a*, *b*, *c*, *d*, *e*, *f*, *g* name the measures, not of the line segments to which they are attached, but rather the measures of the related arcs after projection has taken place. Figures 32a–32c show this very clearly. These drawings give you a simple geometrical construction to be used for arriving at the required arc lengths needed for the layout of the circular bands. The layout is shown in Fig. 32d.

Without entering into too much detail, let it be simply stated that all linear measures of line segments on the pentagon face in Fig. 32 can be calculated from right triangles. Chords of the small circle circumscribing this face, identified as l_1, l_2, and l_3 in Fig. 32, then become chords of a great circle arc as shown in Figs. 32a–32c. This is equivalent to passing a plane through this chord and the center O_3 of the sphere circumscribing the dodecahedron. This produces other sets of right triangles from which the required arc

Fig. 32. A pentagram inscribed in a pentagon face of a dodecahedron. (Figs. 32a–32c. Gnomonic projections.)

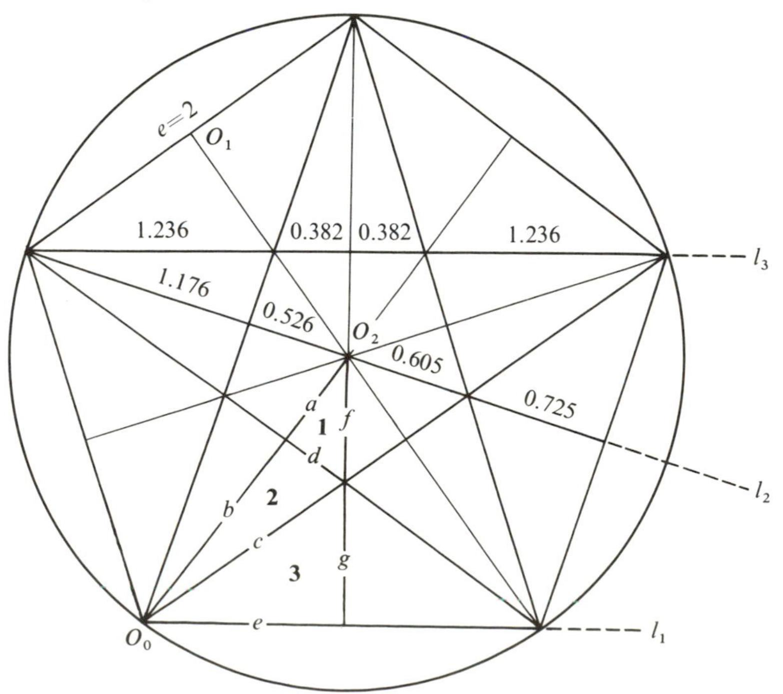

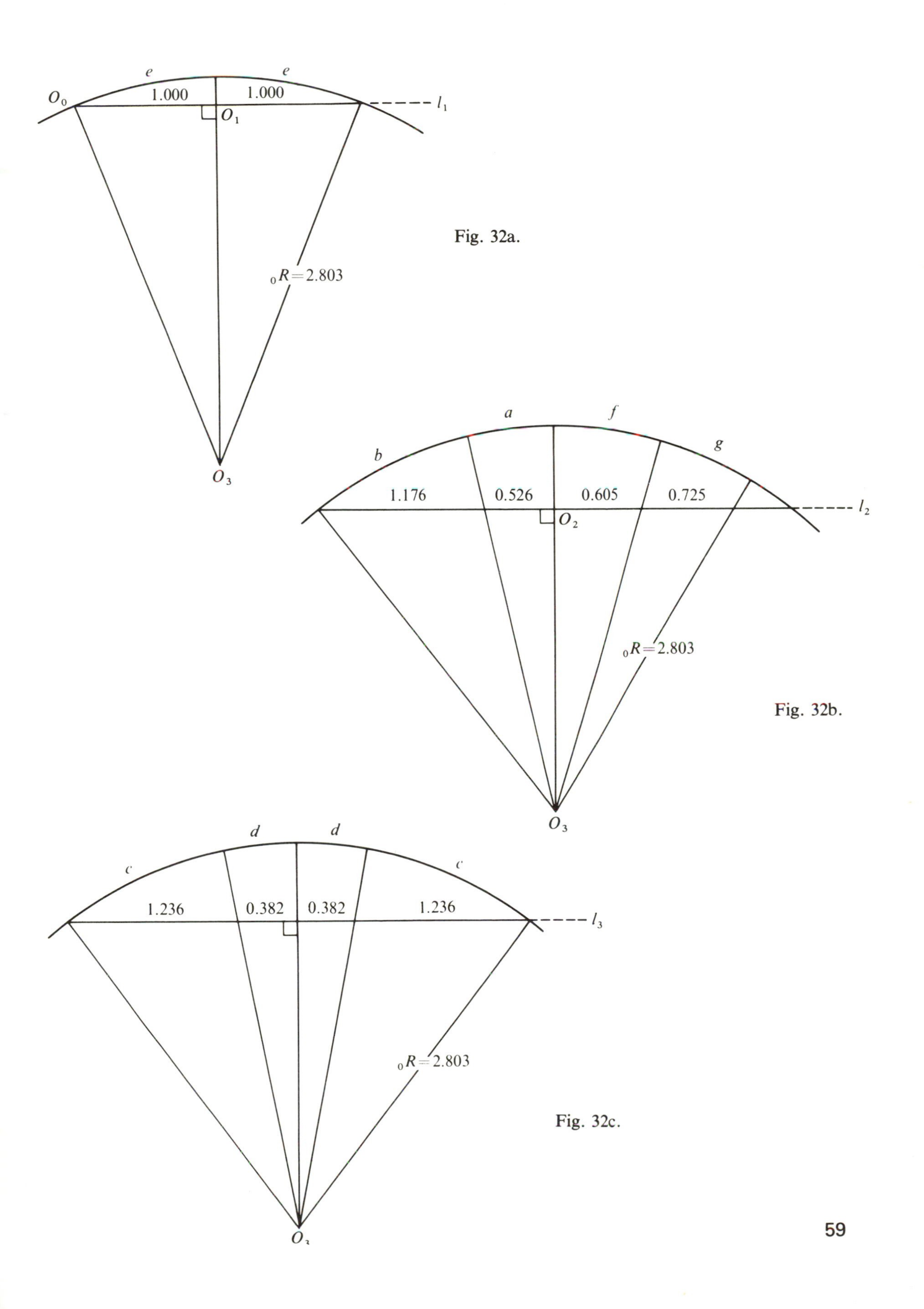

Fig. 32a.

Fig. 32b.

Fig. 32c.

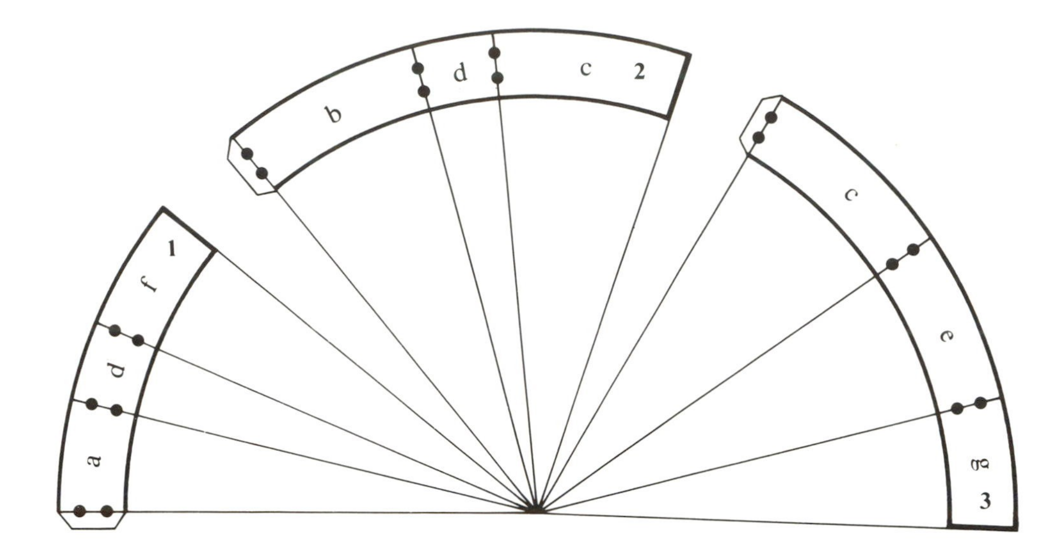

Fig. 32d. Layout of bands for dodecahedron with pentagrams.

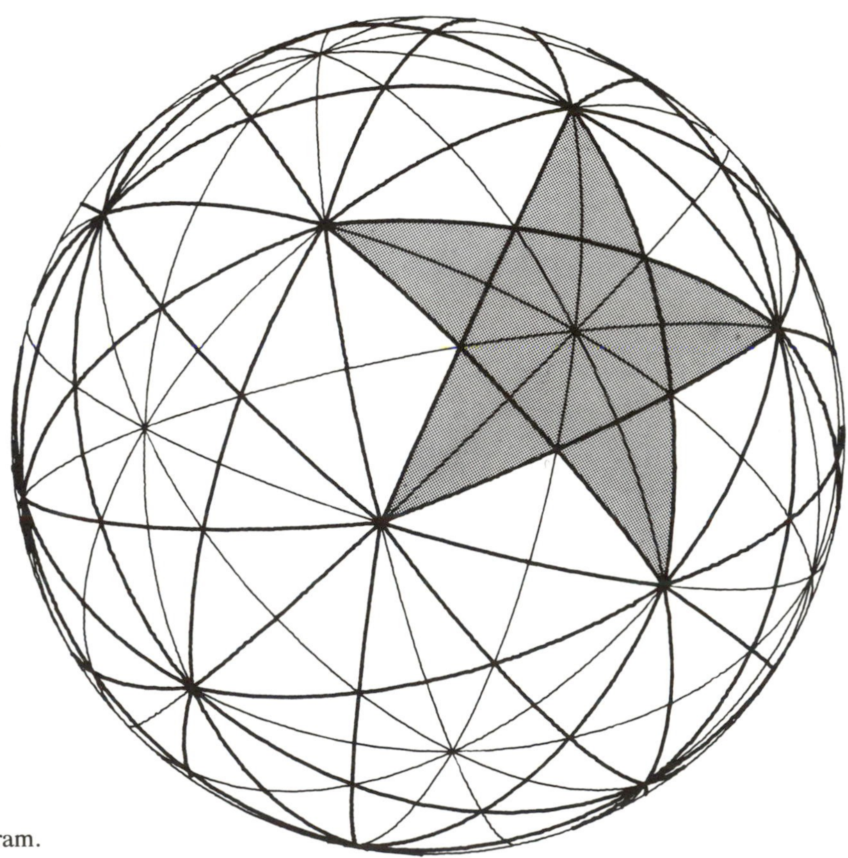

Plate 28. Spherical pentagram.

Photo 21. Dodecahedron with pentagrams.

lengths needed for the bands can be derived.

In Fig. 32 the numerals on line segments are linear measures where an edge length $e = 2$ units. These same numerals appear again in Figs. 32a–32c. To see what arc lengths go with which bands, you must refer again to the pentagon face. Only three kinds of bands are needed. These are named **1** for the inner pentagon, **2** for the star arm, and **3** for the outer pentagon. Each comes in right- and left-handed varieties, as mirror-image pairs.

To make this model, begin by assembling ten parts of **1** for a small pentagon. To this add five star arms, each arm made up of two parts of **2**. Finally, fill the spaces between the star arms with ten parts of **3** to complete the pentagon face with its inscribed star. Twelve such faces complete the model. Plate 28 shows a spherical pentagram, and Photo 21 shows the completed model.

This dodecahedron model reintroduces the right spherical triangle as the basic element of construction. Your feeling may be, and rightly so, doesn't this enormously increase the number of bands and, consequently, the amount of time needed for a complete model? Yes, but if you have perseverance enough, the final result is also enormously attractive.

If you want to make this and subsequent models without worrying about the mathematical calculations, the required numerical data will be given in each case. You are invited to verify the results, however, either before or after making the model.

For the dodecahedron with a pentagram in each face, the arc measures and bands are as follows:

arcs	*bands*
$a = 13.283$	**1** $a\ d\ f$
$b = 24.095$	**2** $b\ d\ c$
$c = 25.788$	**3** $c\ e\ g$
$d = 9.477$	
$e = 20.905$	
$f = 16.267$	
$g = 15.450$	

Icosidodecahedron with its complete set of thirty-one great circles

The icosidodecahedron gives an equally attractive model, if you leave all the triangles with six elements each and treat the pentagons the same as was done for the dodecahedron, that is, with a pentagram inscribed in each. Linear measures remain for the faces, but, because ${}_0R$ changes to that of the icosidodecahedron, all arc lengths must be redrawn. The principles, however, remain the same. Refer back to Fig. 32 for the pentagon face, but then look at Figs. 33a–33c that show how to derive the arcs for the pentagram to be used in this model. Figure 33d gives the triangle face needed here, and Fig. 33e its related arcs.

An interesting feature of this model is that it can be made in such a way that the work can stop after the construction of six pentagons with inscribed pentagrams and their ten intervening equilateral triangles. This gives you a perfectly hemispherical model. But the full spherical model is even more interesting because it contains a total of thirty-one complete great circles, all intersecting in this fantastic spherical tessellation. Photo 22 shows the completed model. The frontispiece shows this model in color.

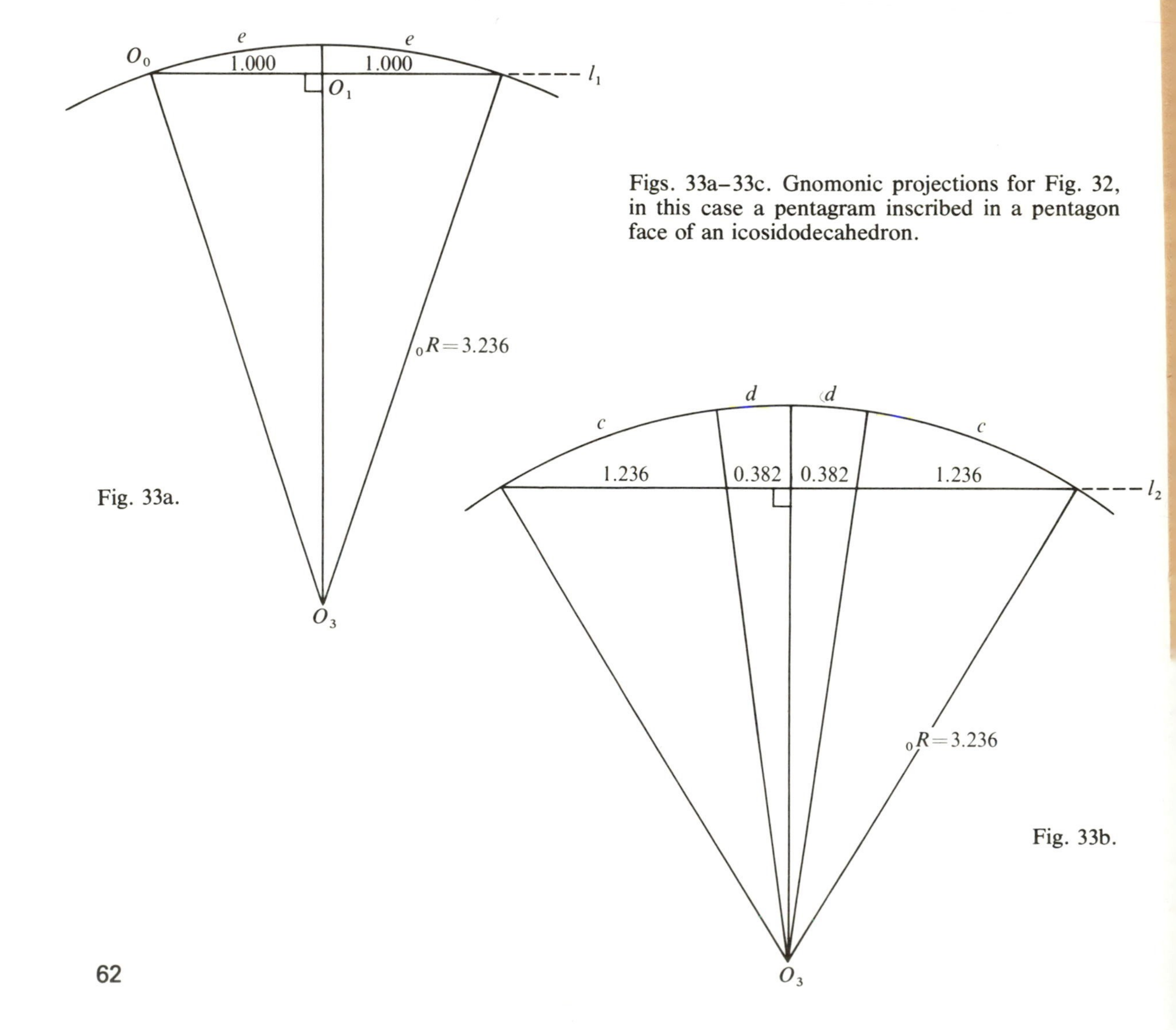

Figs. 33a–33c. Gnomonic projections for Fig. 32, in this case a pentagram inscribed in a pentagon face of an icosidodecahedron.

Fig. 33a.

Fig. 33b.

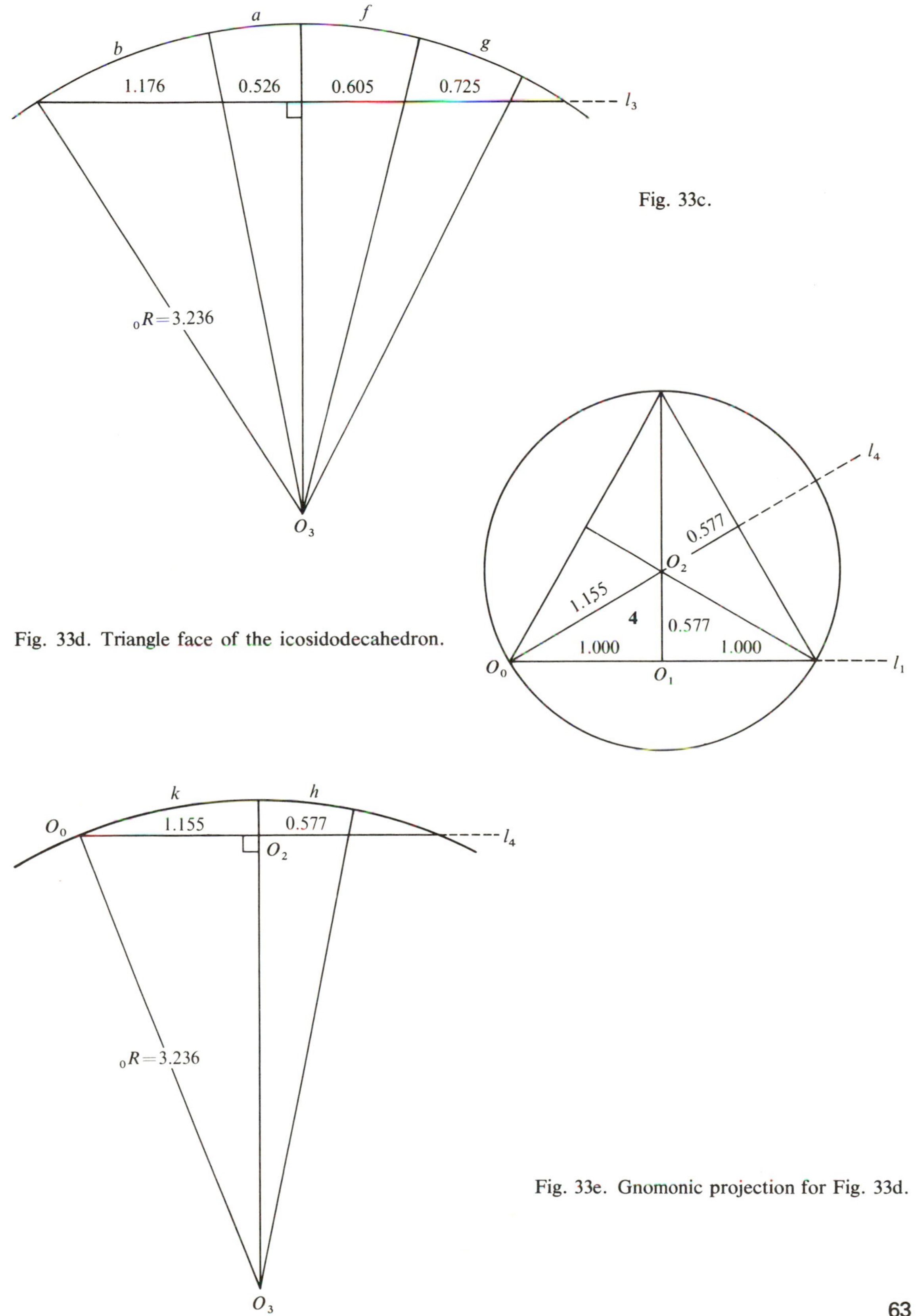

Fig. 33c.

Fig. 33d. Triangle face of the icosidodecahedron.

Fig. 33e. Gnomonic projection for Fig. 33d.

Photo 22. Icosidodecahedron with pentagrams.

For the icosidodecahedron with a pentagram in each pentagon face, the arc measures and bands are as follows:

arcs	*bands*
a = 10.812	**1** $a\ d\ f$
b = 20.905	**2** $b\ d\ c$
c = 22.239	**3** $c\ e\ g$
d = 7.761	**4** $h\ e\ k$
e = 18.000	
f = 13.283	
g = 18.434	
h = 10.812	
k = 20.905	

Each band forms a right spherical triangle. Ten of each of **1, 2, 3** make a spherical pentagon with its inscribed pentagram; six of **4** make an equilateral triangle.

Simpler models, omitting a and r or e, a, and r are shown in Photos 23 and 24.

Photo 23. Icosidodecahedron with pentagrams (omitting a and r).

Photo 24. Icosidodecahedron with pentagrams (omitting e, a, and r).

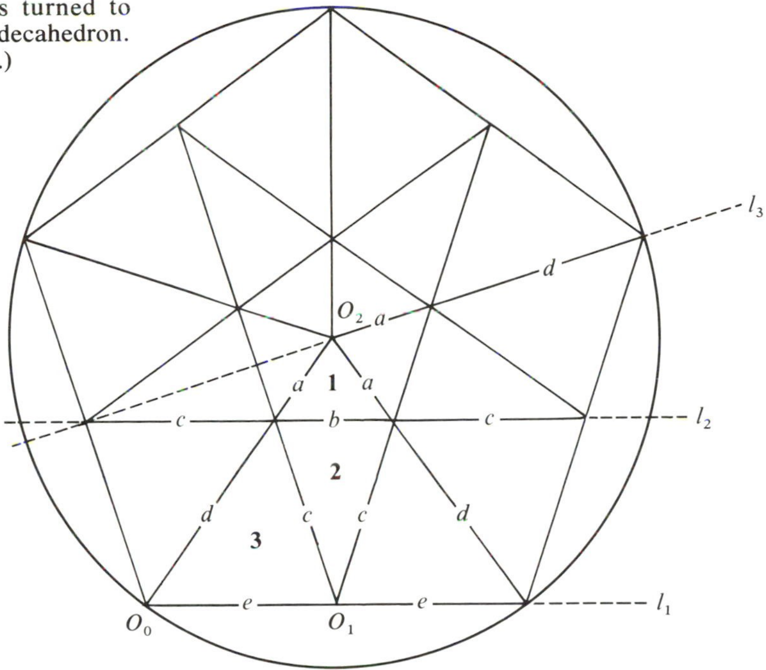

Fig. 34. A pentagram with vertices turned to midedge in a pentagon face of a dodecahedron. (Figs. 34a–34c. Gnomonic projections.)

Dodecahedron with pentagrams turned

If the stars on the faces of the dodecahedron are turned so that their vertices are at the midpoints of the edges, another very interesting model emerges. The drawing of one face is shown in Fig. 34 with the projections in Figs. 34a–34c.

For this model, the arc measures and bands are as follows:

arcs	*bands*
$a = 13.283$	**1** $a\ b\ a$
$b = 15.522$	**2** $c\ b\ c$
$c = 22.239$	**3** $c\ e\ d$
$d = 24.095$	
$e = 20.905$	

The way is now open for you to treat other polyhedrons having pentagon faces in the same way. Variations omitting the apothem can also lead to very successful models. Omitting radial lines is less successful.

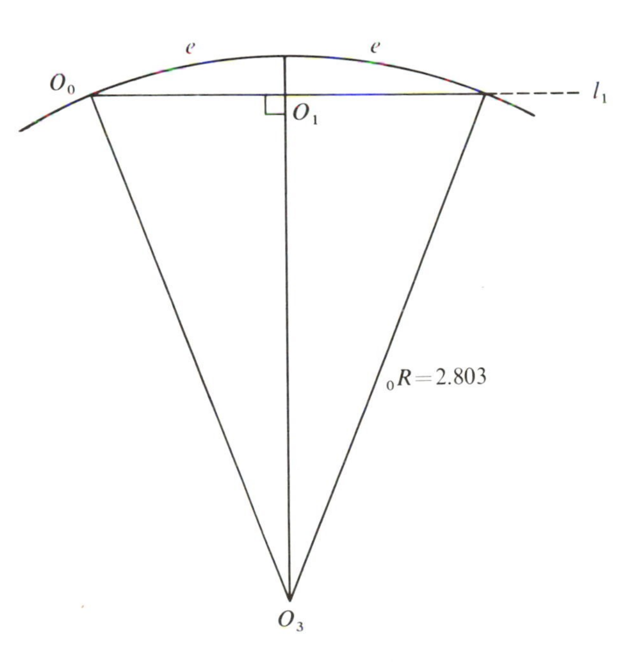

Fig. 34a.

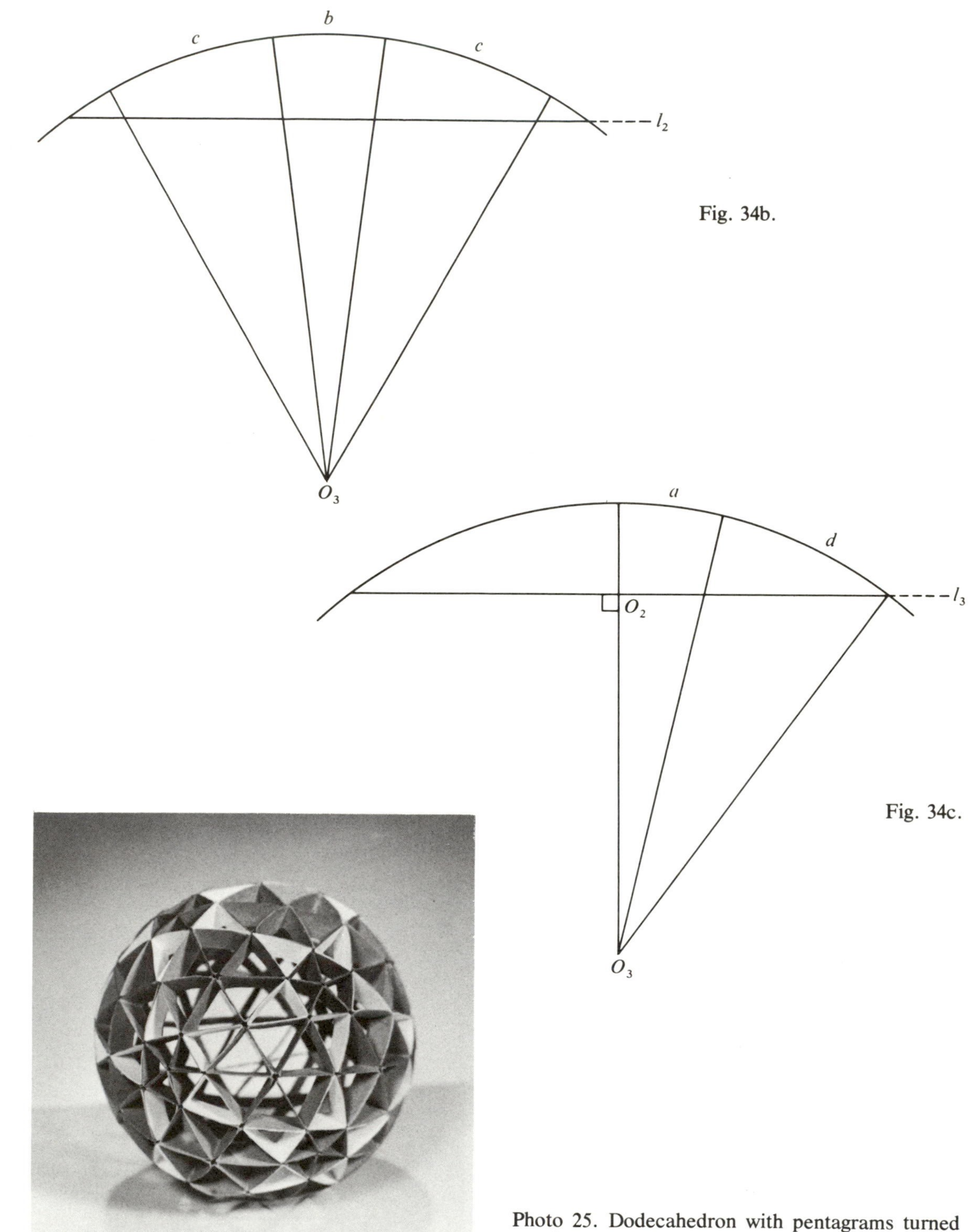

Fig. 34b.

Fig. 34c.

Photo 25. Dodecahedron with pentagrams turned to midedge.

Other star-faced models

Then proceeding further, the hexagram can be inscribed in a hexagon face, the octagram in an octagon face, and the decagram in a decagon face. These facial planes are illustrated in Figs. 35a–35c. Once the linear measures have been calculated for these stars, they can be used repeatedly to calculate the arc measures required for each different model. (See Figs. 36–39, which give a purely geometrical approach to the models.) The calculation of arc lengths will be left for you to do as an exercise.

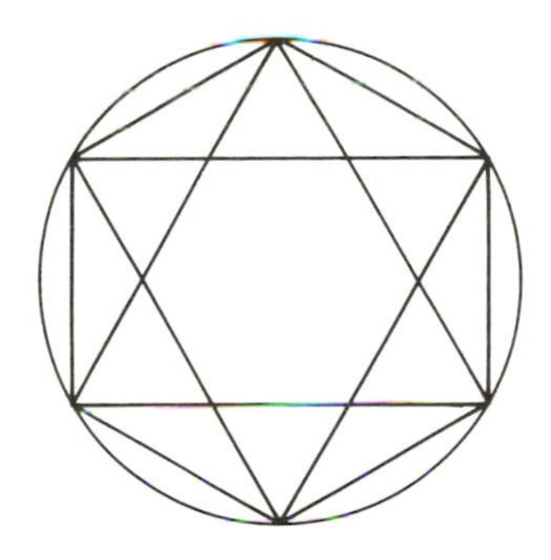

Fig. 35a. A hexagram inscribed in a hexagon.

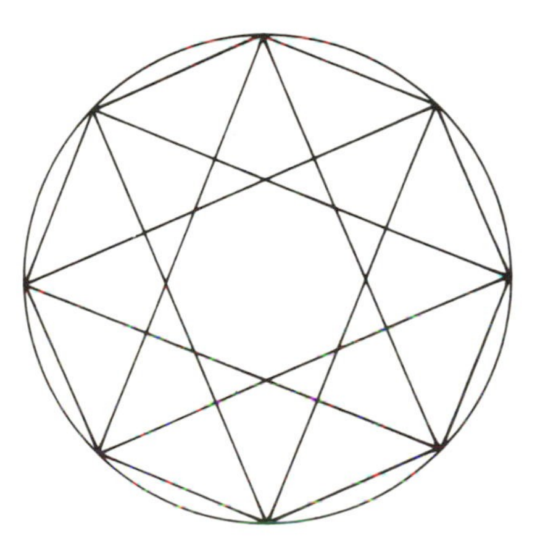

Fig. 35b. An octagram inscribed in an octagon.

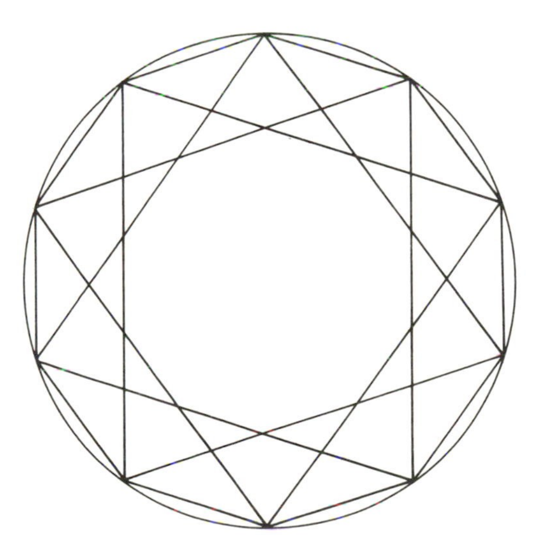

Fig. 35c. A decagram inscribed in a decagon.

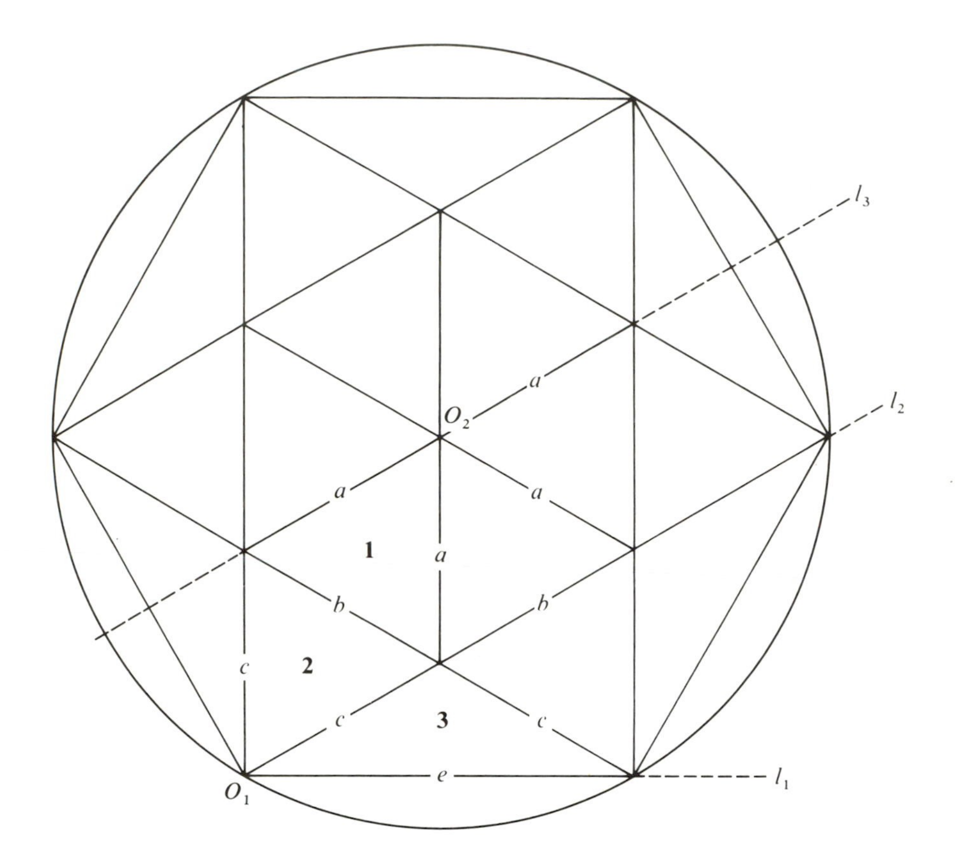

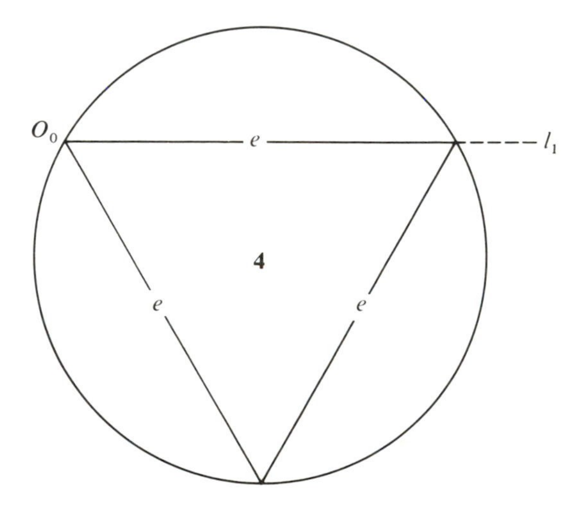

Fig. 36. Facial planes of the truncated tetrahedron with a hexagram inscribed in the hexagon face. (Figs. 36a–36c. Gnomonic projections.)

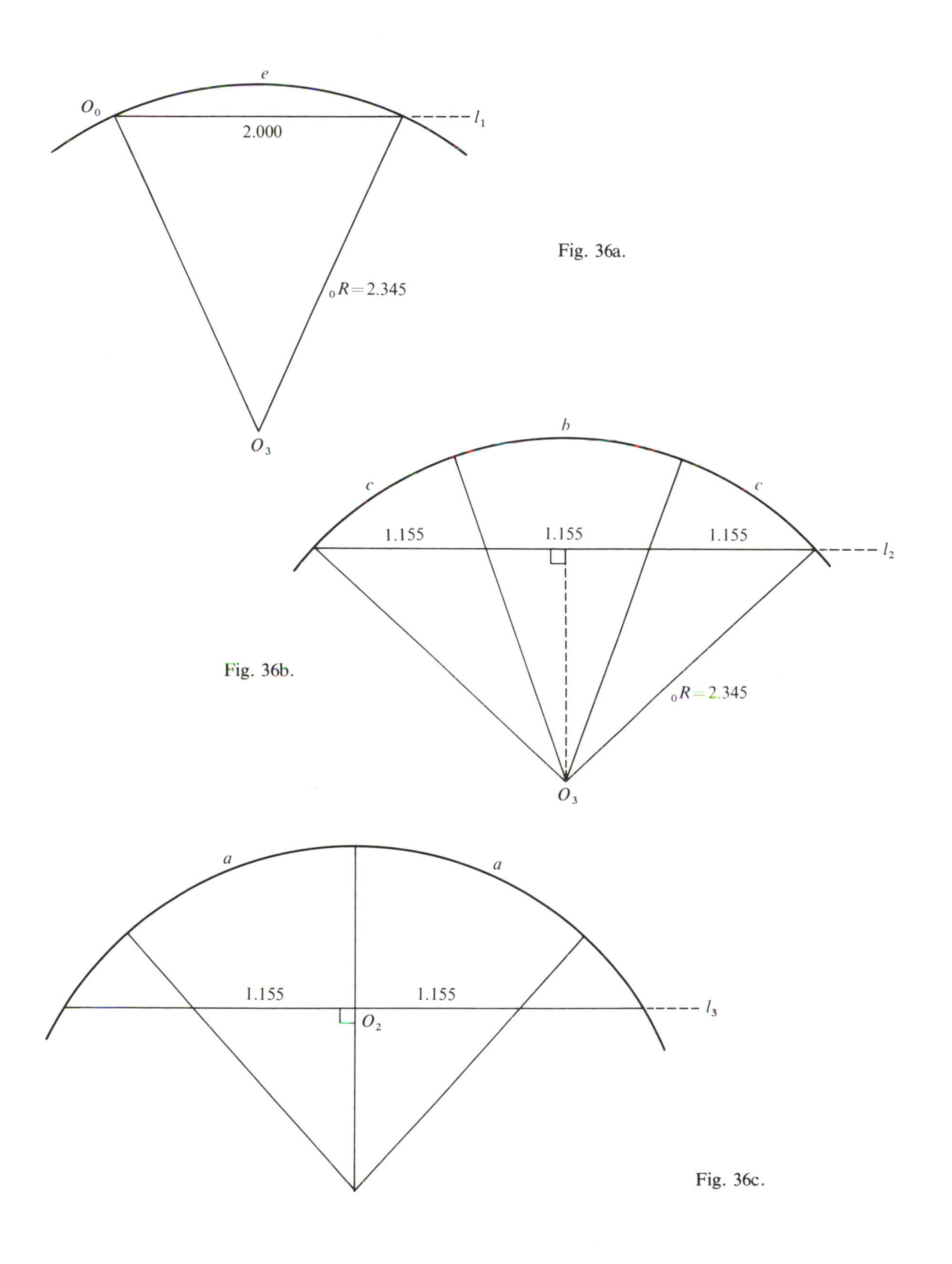

Fig. 36a.

Fig. 36b.

Fig. 36c.

Fig. 37. Facial planes of the truncated cube with an octagram inscribed in the octagon face. (Figs. 37a–37c. Gnomonic projections.)

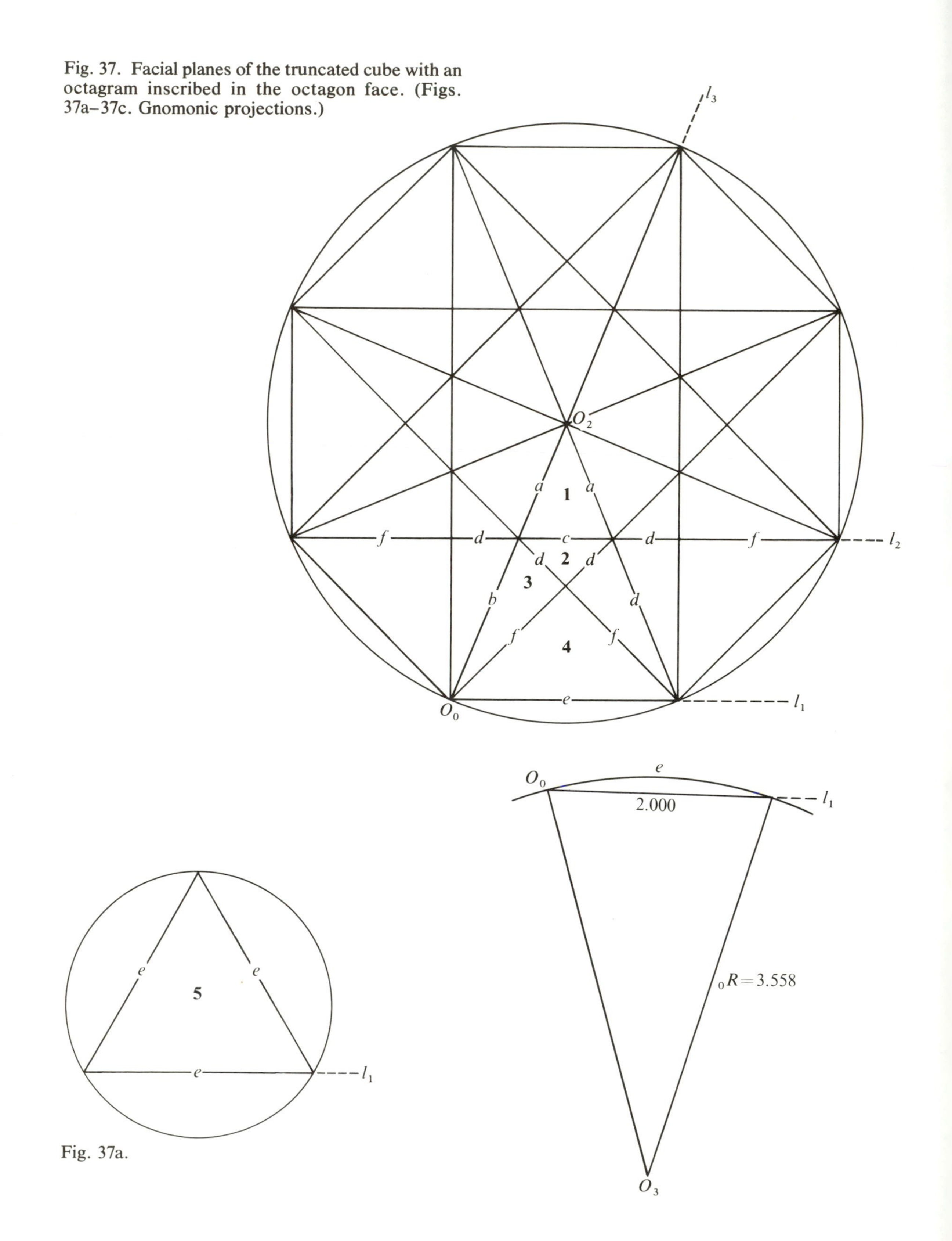

Fig. 37a.

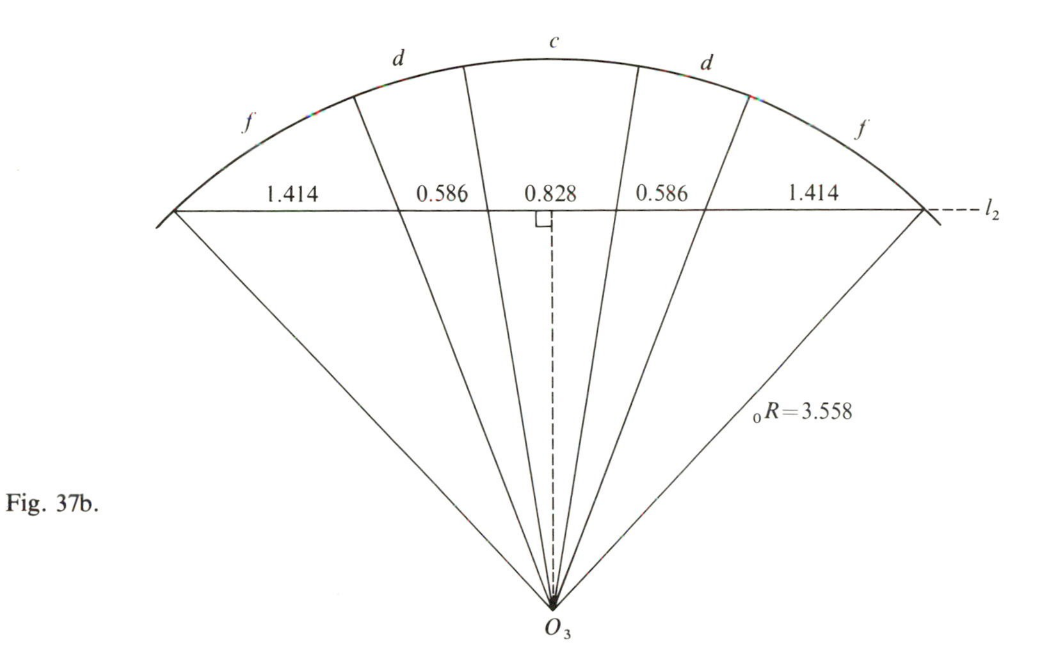

Fig. 37b.

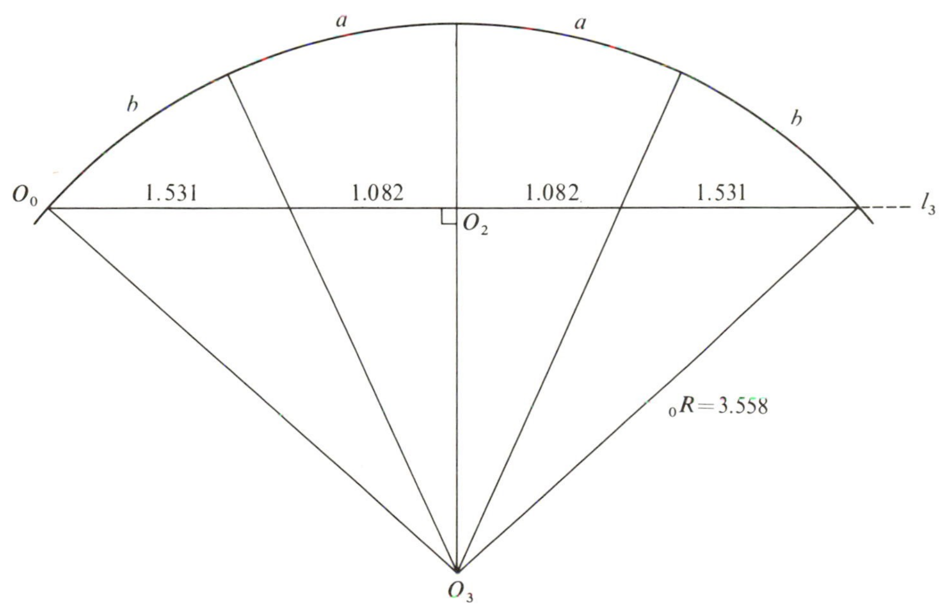

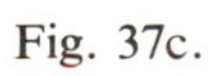
Fig. 37c.

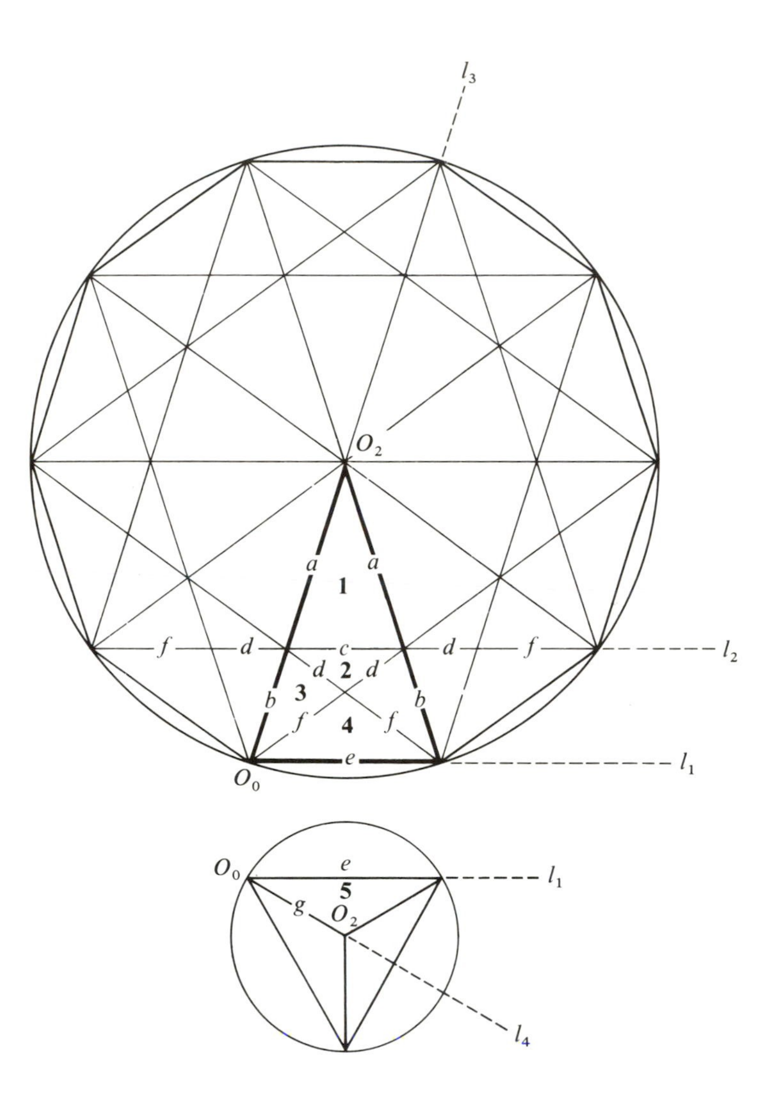

Fig. 38. Facial planes of the truncated dodecahedron with a decagram inscribed in the decagon face. (Figs. 38a–38c. Gnomonic projections.)

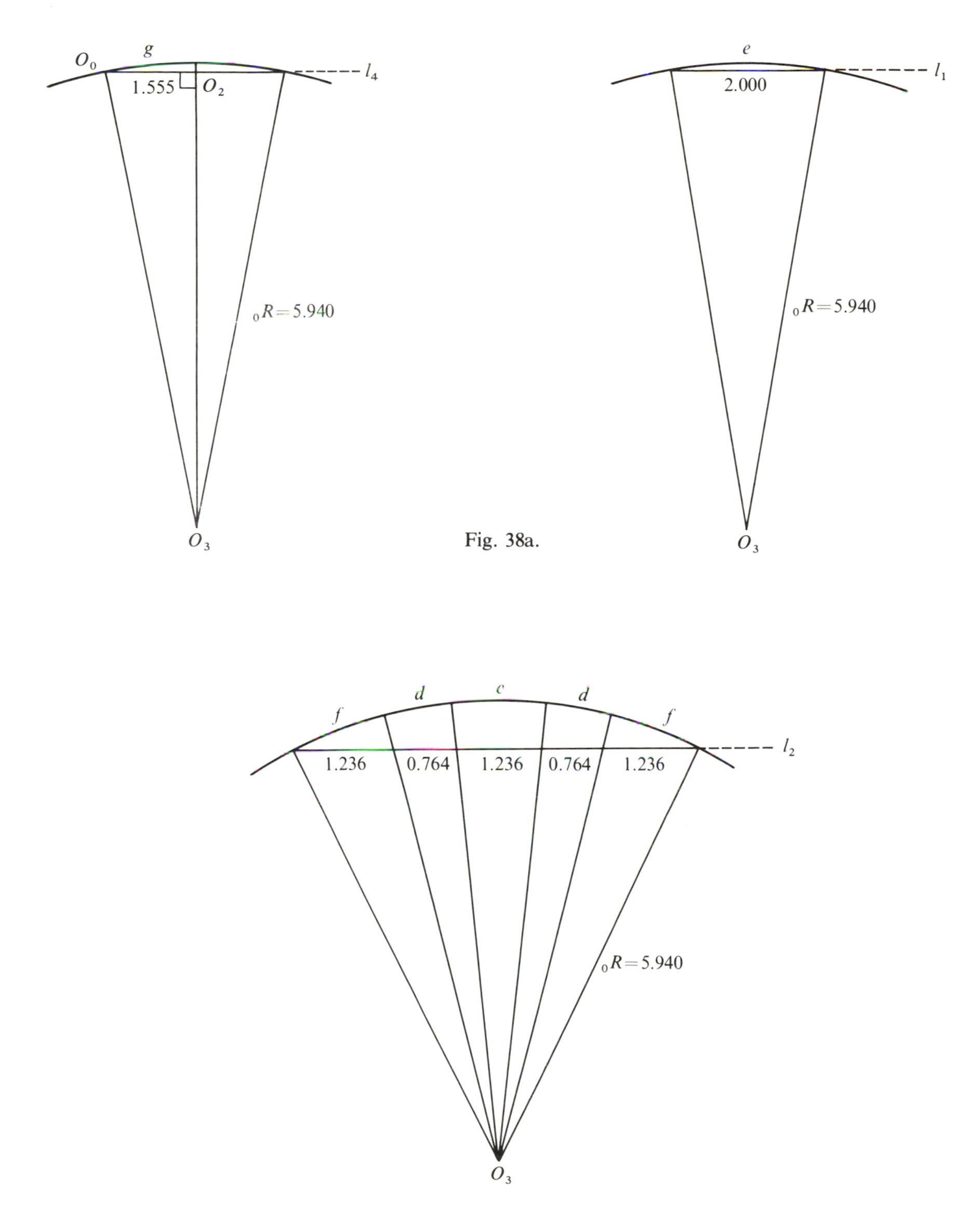

Fig. 38a.

Fig. 38b.

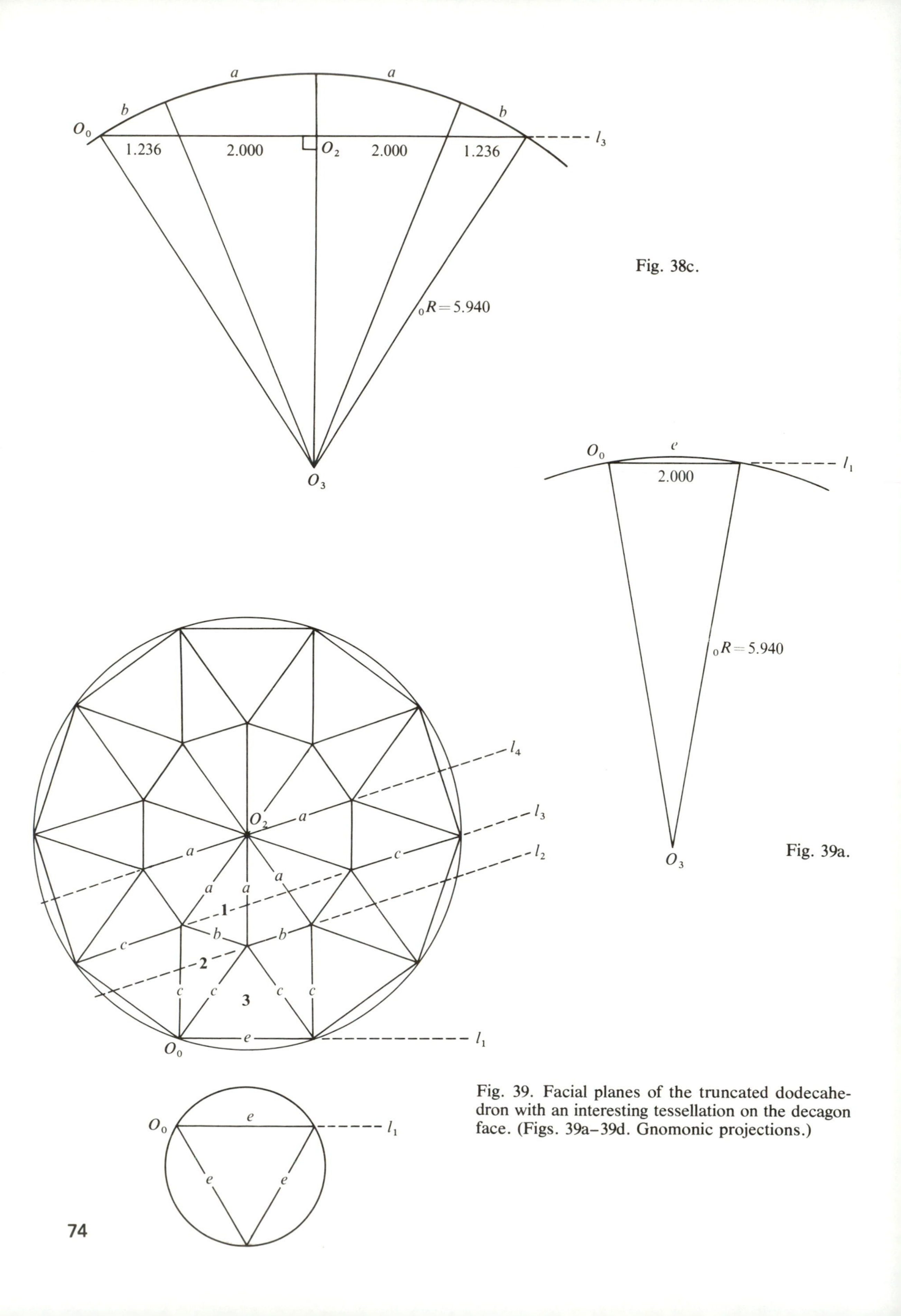

Fig. 39. Facial planes of the truncated dodecahedron with an interesting tessellation on the decagon face. (Figs. 39a–39d. Gnomonic projections.)

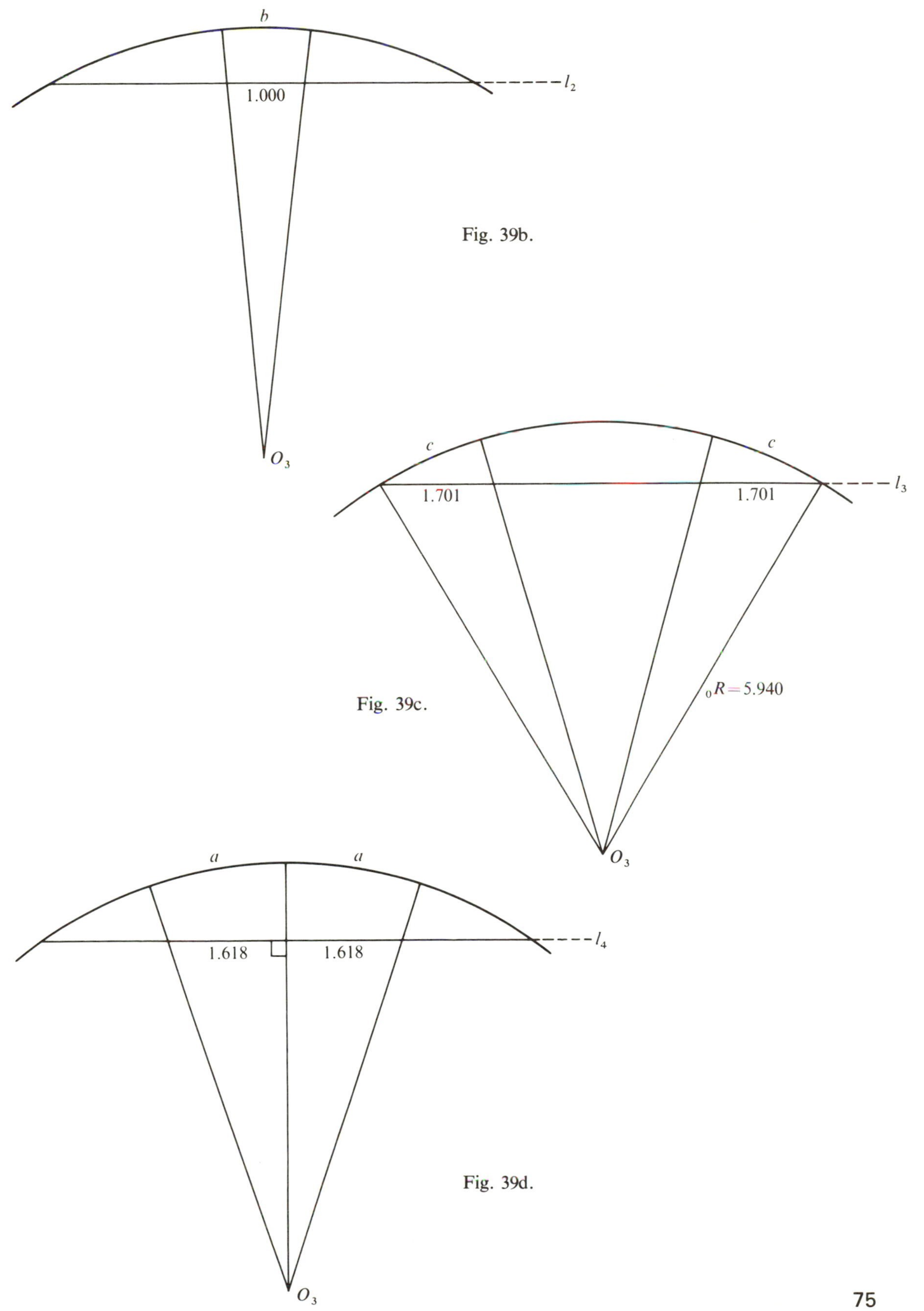

Fig. 39b.

Fig. 39c.

Fig. 39d.

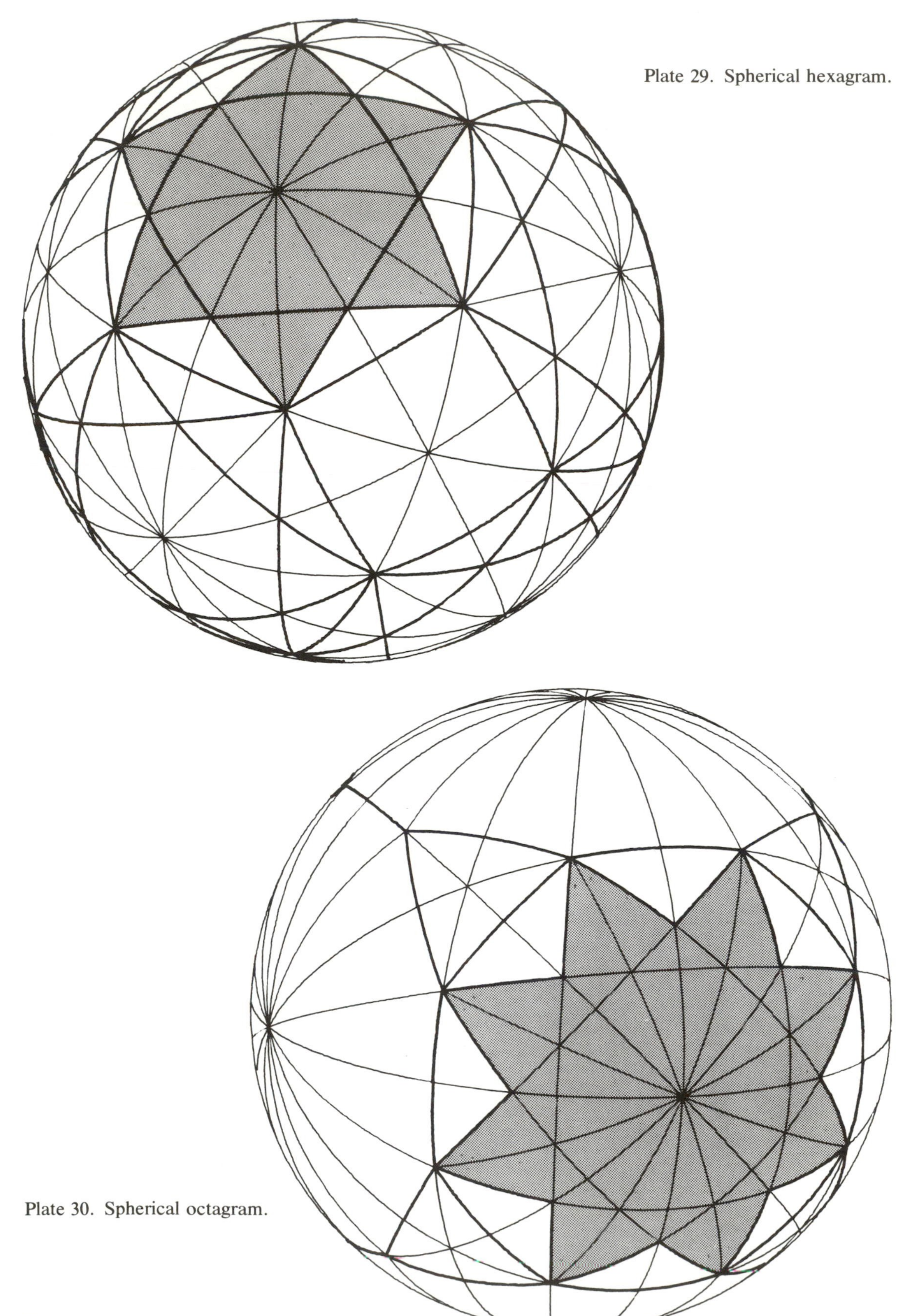

Plate 29. Spherical hexagram.

Plate 30. Spherical octagram.

Plate 29 shows a spherical hexagram, and Plate 30 shows a spherical octagram. Photos 26–32 illustrate some models done according to the methods set forth here. Each is identified so that you can have some idea of its appearance should you want to make a model of it for yourself.

After you have done a few of these, you will see their close relationship to some of the nonconvex uniform polyhedrons. No attempt will be made here to give a detailed description or explanation of this relationship. Some further ideas will be presented in Section V of this book, where polyhedral density will be discussed. The study of this topic will be left as a tantalizing area for your own investigation.

Photo 26. Truncated tetrahedron with hexagrams (omitting *r*) and triangles (omitting *a* and *r*).

Photo 27. Truncated tetrahedron with hexagrams (omitting *e* and *a*) and triangles (omitting *e*).

Photo 28. Truncated cube with a special tessellation on the octagon face.

Photo 29. Truncated cube with octagrams (omitting a) and triangles (omitting a and r).

Photo 30. Truncated dodecahedron with decagrams (omitting a).

Photo 31. Truncated dodecahedron with an interesting tessellation on the decagon face.

Photo 32. Rhombicosidodecahedron with pentagrams and squares (omitting a) and triangles (omitting a and r).

IV. Geodesic domes

If you have ever had the opportunity of looking closely at a geodesic dome, your first impression may well have been that all its triangles are equilateral. A second impression may well have been that these triangles seem to group themselves into hexagons, six around a point. Closer inspection may have revealed to you that some groups are not hexagons but rather pentagons. So obviously these triangles at least cannot be equilateral.

It is now time to enter more deeply into a study of geodesic domes. What you have done so far in making models of spherical polyhedrons will serve as a good background for this investigation.

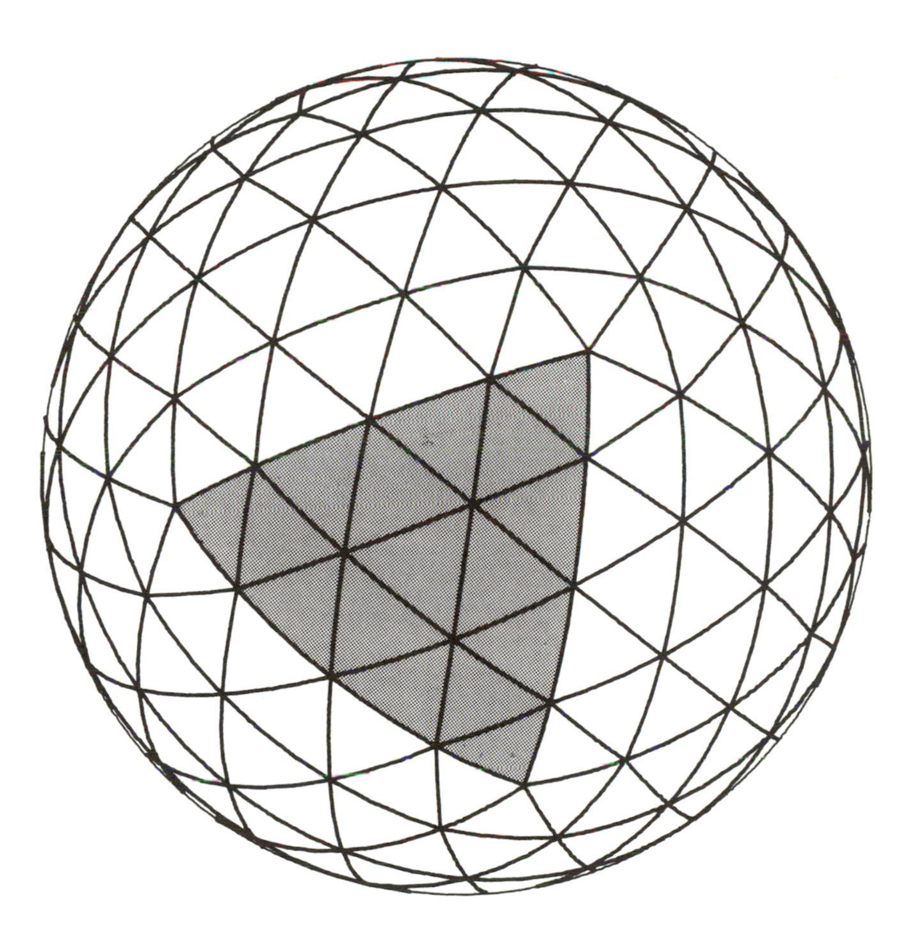

Plate 31. A 4-frequency icosahedral geodesic dome. $\{3, 5+\}_{4,0}$

Photo 33. Dodecahedron (omitting *a*) or pentakis-dodecahedron. $\{3, 5+\}_{1,1}$

Moreover the procedure for making spherical models, as it has been developed in this book, can easily be extended to making models of geodesic domes in paper. For this purpose it is good to return to the regular and semiregular models because some variations here will provide an easy introduction to this topic.

The simplest geodesic domes

Since there are three regular spherical models, the tetrahedral, octahedral, and icosahedral, you may already have made the variations of these that eliminated *a* and *r*. The tetrahedron made in this way has four equilateral triangles, the octahedron has eight, and the icosahedron has twenty. These are in fact the simplest examples of geodesic domes. They do not look very much like geodesic domes as generally known however. The spherical dodecahedron will be the first one to take on such as appearance, provided you make it retaining *r* and eliminating only *a*. This polyhedron is more correctly called a pentakisdodecahedron. The name denotes the number of its faces: $5 \times 12 = 60$. Here each face of the dodecahedron has been decomposed into a set of five isosceles triangles. Use the circular band shown in Fig. 40 to make this model; see also Photo 33.

Once the model is made examine it and you will see that it has triangles that your eye can arrange into groups of six or into groups of five. The incenters of these groups are called hexavalent and pentavalent vertices. This is an important feature of geodesic domes as generally known. You will find references to this feature again and again in what follows. The pentakisdodecahedron model has twelve pentavalent vertices and twenty hexavalent ones.

The icosidodecahedron becomes an example of a geodesic dome if all its triangles are allowed to remain as equilateral and the pentagons are changed into groups of five isosceles triangles each. See Fig. 41 for the layout of the bands. Photo 34 shows the complete model. This model has twelve pentavalent vertices and thirty hexavalent ones. Notice too that as

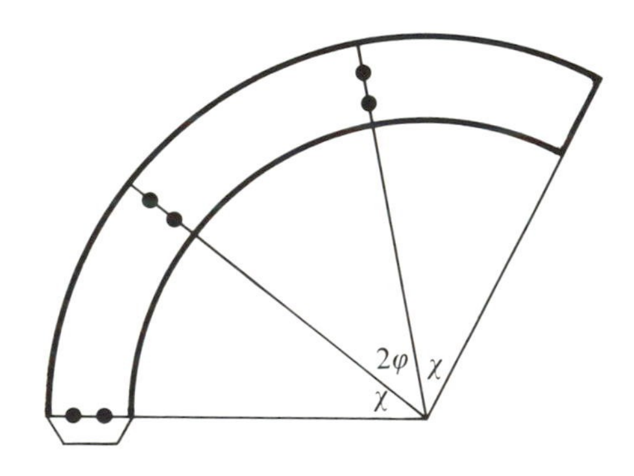

Fig. 40. Band for a spherical dodecahedron (omitting *a*) or for a pentakisdodecahedron.

the number of elements increases the models become more attractive.

The truncated icosahedron has twelve pentagon faces and twenty hexagon faces. If the pentagons are constructed with five elements each and the hexagons with six elements each, another model of a genuine geodesic dome results. See Fig. 42 for the layout of the bands and Photo 35 for the complete model. This model has twelve pentavalent vertices and eighty hexavalent ones. Not only has the number of elements grown larger here, but the number of hexavalent vertices as well. Notice that the number of pentavalent vertices remains constant at twelve. If you find it difficult to verify the count of hexavalent vertices in this case, do not worry about it now. Later on a formula will be given from which this and even larger counts will easily follow.

You must have noticed that the arc lengths for the bands are to be derived from Tables 1 and 2 at the end of Section I and Table 3 at the end of Section II. This means that these models are genuine geodesic domes that have all been made from circular bands of paper derived by ruler and compass construction alone and their arc lengths can be calculated by ordinary trigonometry.

There is one more of the semiregular

Photo 34. Icosidodecahedron with pentagons (omitting a) and triangles (omitting a and r). $\{3, 5+\}_{2,0}$

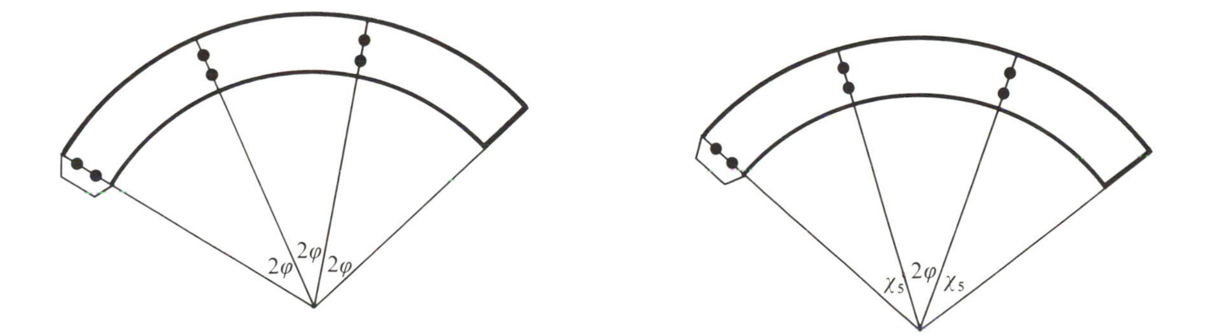

Fig. 41. Bands for an icosidodecahedron.

Photo 35. Truncated icosahedron with pentagons (omitting a) and hexagons (omitting a). $\{3, 5+\}_{3,0}$

polyhedrons for which this can be done. It is the snub dodecahedron: See Fig. 43 for the layout of the bands and Photo 36 for the completed model. It has twelve pentavalent vertices and sixty hexavalent ones. Because it is derived from the snub dodecahedron, you might also suspect that it can be done in right- or left-handed varieties. Its twisted symmetry also makes it somewhat different from the others. Precisely what this implies will be seen later.

Geodesic domes derived from the icosahedron

It will now be shown how you can proceed to the construction of more complex

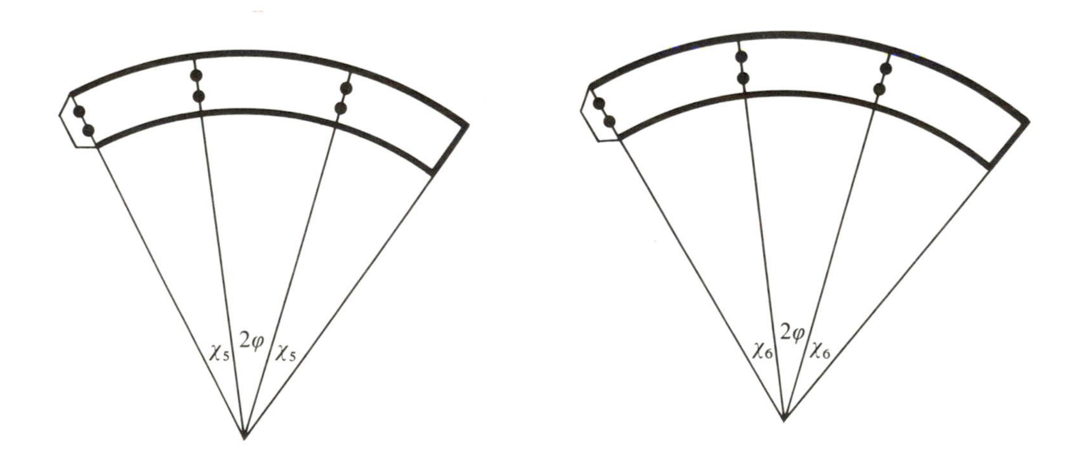

Fig. 42. Bands for a truncated icosahedron.

domes. It is easiest to begin with the icosahedron. The idea here is to decompose one equilateral triangle face of the icosahedron into a tessellated network of smaller triangles. From plane geometry you known that the line joining the midpoints of two sides of a given triangle is parallel to its third side and equal to half of it. If the given triangle is equilateral to begin with, then joining respective midpoints yields a smaller equilateral triangle whose vertices lie on the sides of the larger given one. The sides of this smaller triangle therefore have linear measures equal to half those of the sides of the larger triangle. In particular, if the edges of the larger triangle are each 2 units in length, the smaller has edges that are 1 unit long.

Photo 36. Snub dodecahedron with pentagons (omitting *a*) and triangles (omitting *a* and *r*). $\{3, 5+\}_{2,1}$

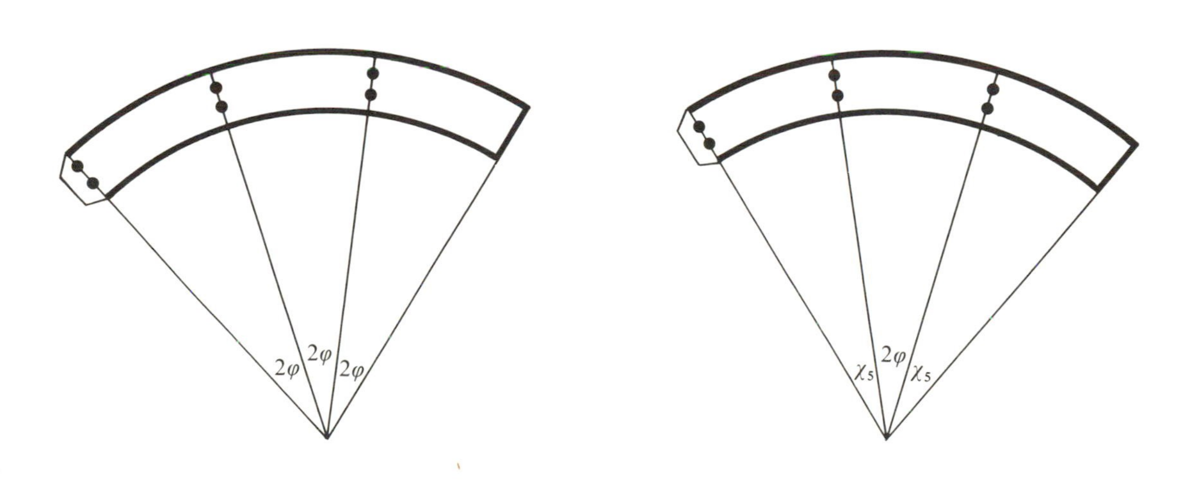

Fig. 43. Bands for a snub dodecahedron.

A 2-frequency model

Suppose now this tessellated network of triangles is drawn on each face of a regular icosahedron and then projected by central or gnomonic projection onto the surface of its circumscribing sphere (see Fig. 44). The method used in Section III for projecting pentagrams and other stars from the flat faces of the respective polyhedrons onto the circumscribing spheres is still valid here.

You should have no trouble interpreting the drawings of Figs. 44a and 44b and calculating the arc lengths. For this model, the arc measures and bands are as follows:

arcs	*bands*
$a = 31.717$	**1** $a\ b\ a$
$b = 36.000$	**2** $b\ b\ b$

It might surprise you to notice that the bands needed for this model are exactly the same as those used for the icosidodecahedron shown in Fig. 41 and in Photo 34. It just happens to be identically the same model derived in two different ways, using basically two different polyhedrons. But you may, perhaps, have been disturbed by the fact that the hexavalent vertices in this model are the incenters of hexagons that are at best only semiregular, being rather long across one axis of symmetry as compared to the other. This situation however improves as the number of elements becomes larger in more complex geodesic domes.

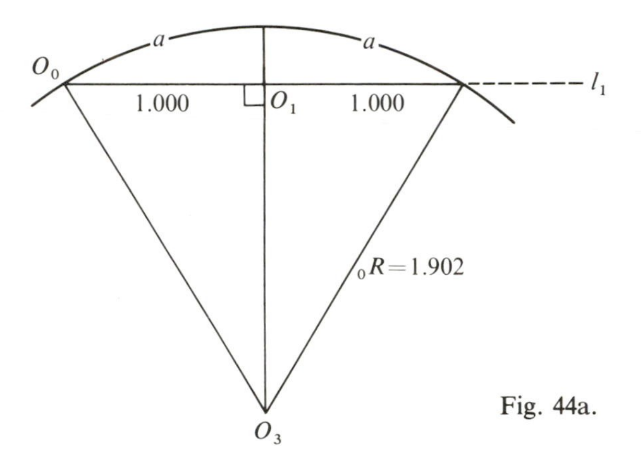

Fig. 44a.

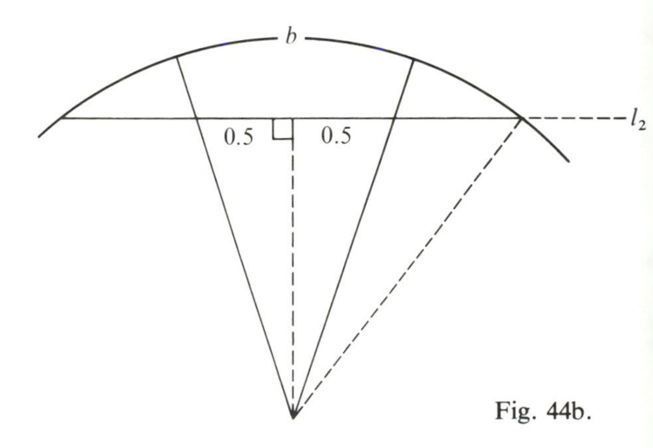

Fig. 44b.

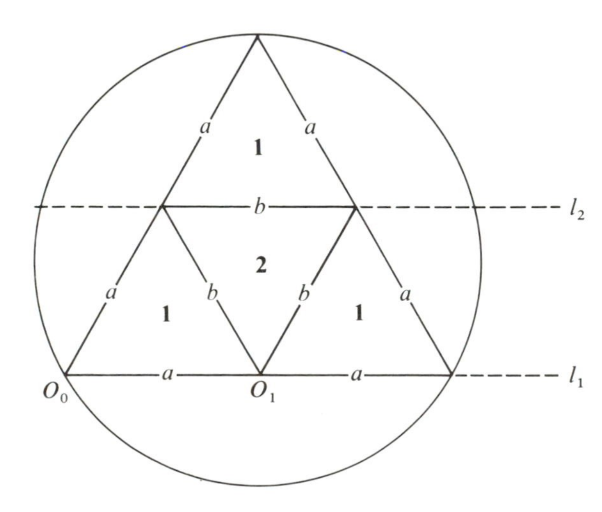

Fig. 44. A 2-frequency icosahedron. (Figs. 44a, 44b. Gnomonic projections.)

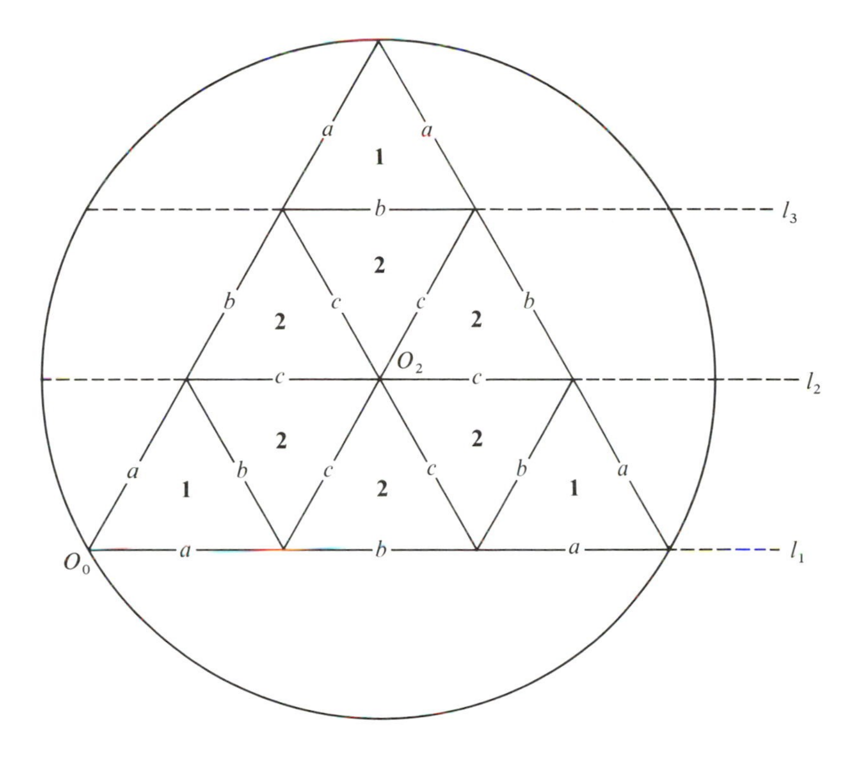

Fig. 45. A 3-frequency icosahedron. (Figs. 45a–45c. Gnomonic projections.)

A 3-frequency model

So having divided the icosahedral face into a tessellated network of four smaller triangles, the next case to try is obviously that suggested by Fig. 45. This is the icosahedral face with each edge broken into three equal segments, giving a total of nine smaller triangles. Figures 45a–45c show how this tessellated network can be gnomonically projected onto a circumscribing sphere. Three different planes through the center of the sphere must be considered. Notice that chord l_3 is equal in length to chord l_1. Hence the central linear segment of l_1 is equal to the segment on l_3. Thus both have the arc measure b.

For the 3-frequency model, the arc measures and bands are as follows:

arcs	*bands*
$a = 20.076$	**1** $a\ b\ a$
$b = 23.281$	**2** $c\ b\ c$
$c = 23.800$	

The completed model is shown in Photo 37. Notice however that this model is the same as the truncated icosahedron shown in Fig. 42 and in Photo 35. This is also an example of a geodesic dome derived from two different polyhedrons, both producing the same model. On the other hand, this fact should not be too surprising because the polyhedrons really both belong to the same symmetry group mathematically speaking.

Photo 37. A 3-frequency icosahedron. $\{3, 5+\}_{3,0}$

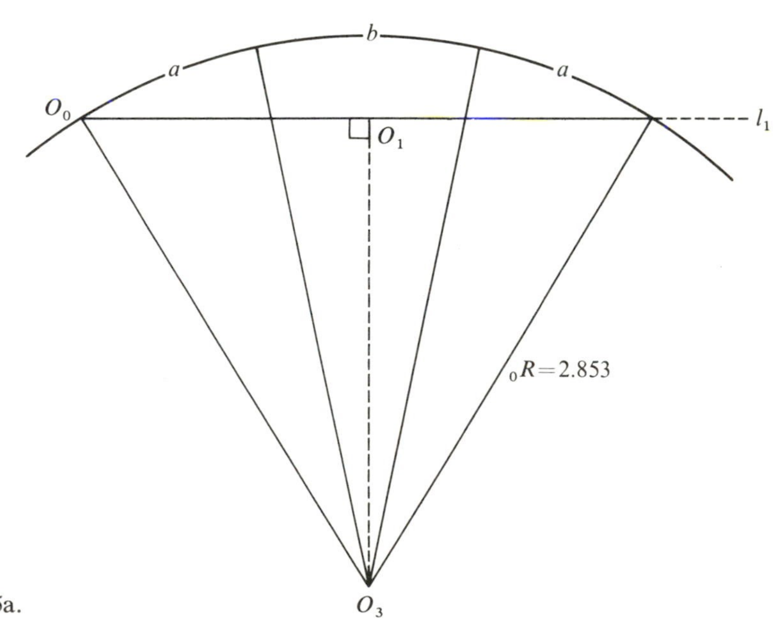

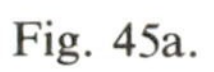

Fig. 45a.

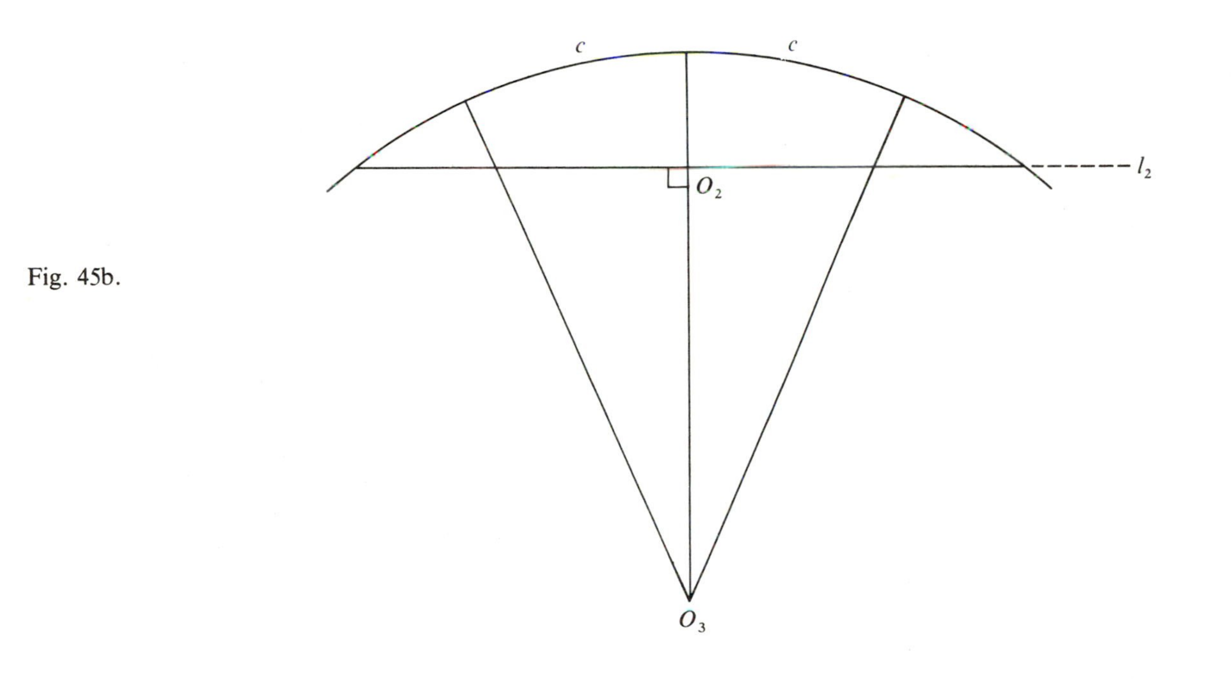

Fig. 45b.

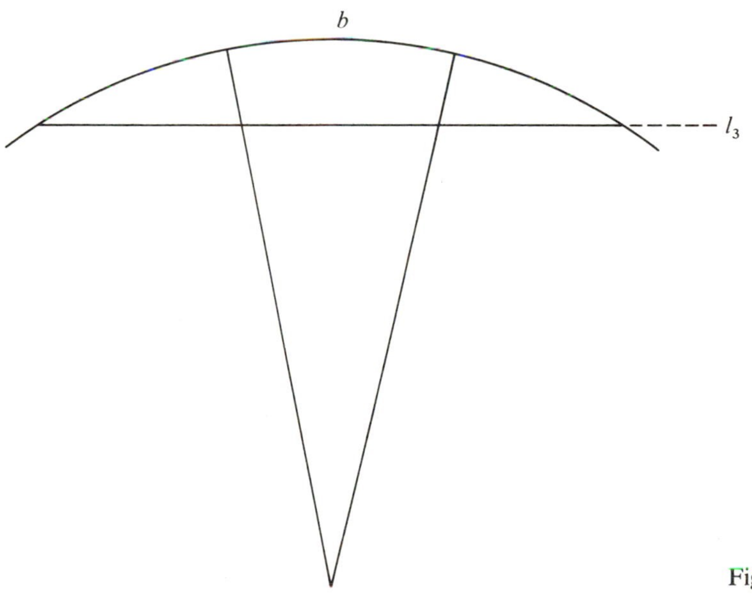

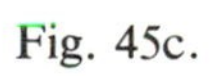

Fig. 45c.

Higher-frequency models

The icosahedron face has now been decomposed into sets of four and nine smaller triangles. You can see that there is no intrinsic reason why this process may not be continued. You may perhaps be delighted to know that 4-, 5-, 6-, and even 8-frequency segmentation can be done using the same methods as those used for the 2-frequency and 3-frequency models. Geometrical constructions become more numerous and calculations as well, but the process remains the same.

Figure 46 shows the icosahedron face in a 4-frequency segmentation with one further feature added, namely the chord l_5. This happens to be a diameter of the small circle circumscribing the icosahedron face. It is shown in its transferred position in Fig. 46e. The linear segments marked k, for one of which $k = r + a$, define end points for perpendiculars p_1, p_2, p_3, p_4 (these occur in Figs. 46a, 46b,

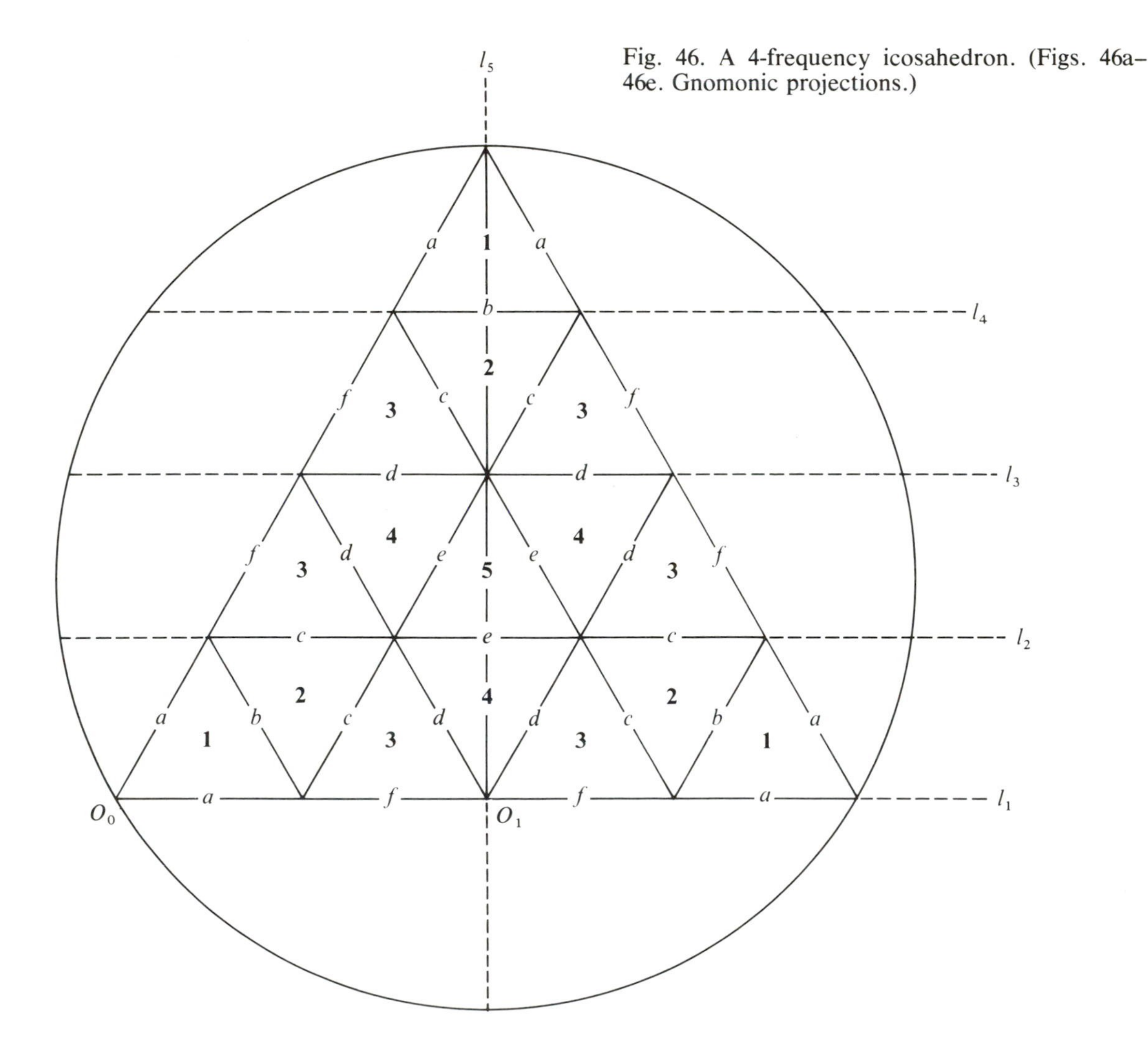

Fig. 46. A 4-frequency icosahedron. (Figs. 46a–46e. Gnomonic projections.)

46c, 46d, respectively). The lengths of these perpendiculars can be found by ordinary trigonometry applied to Fig. 46e, since a sufficient number of right-angled triangles is found here. Once these perpendiculars have been calculated, their measures are used in the other drawings to calculate the arc lengths *a, b, c, d, e, f*.

In this 4-frequency icosahedron, the arc measures and bands are as follows:

arcs	*bands*	
$a = 14.545$	**1**	*a b a*
$b = 16.978$	**2**	*c b c*
$c = 16.937$	**3**	*c d f*
$d = 18.000$	**4**	*d e d*
$e = 18.699$	**5**	*e e e*
$f = 17.172$		

Photo 38. $\{3, 5+\}_{4,0}$

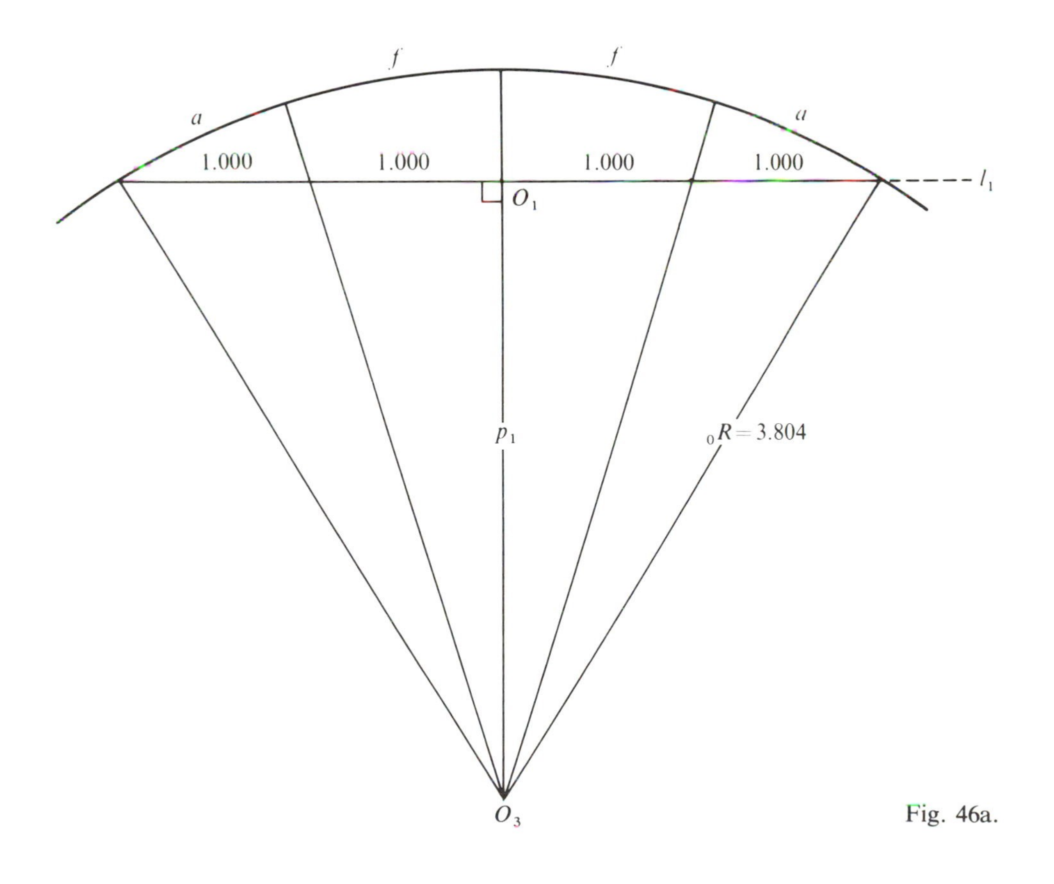

Fig. 46a.

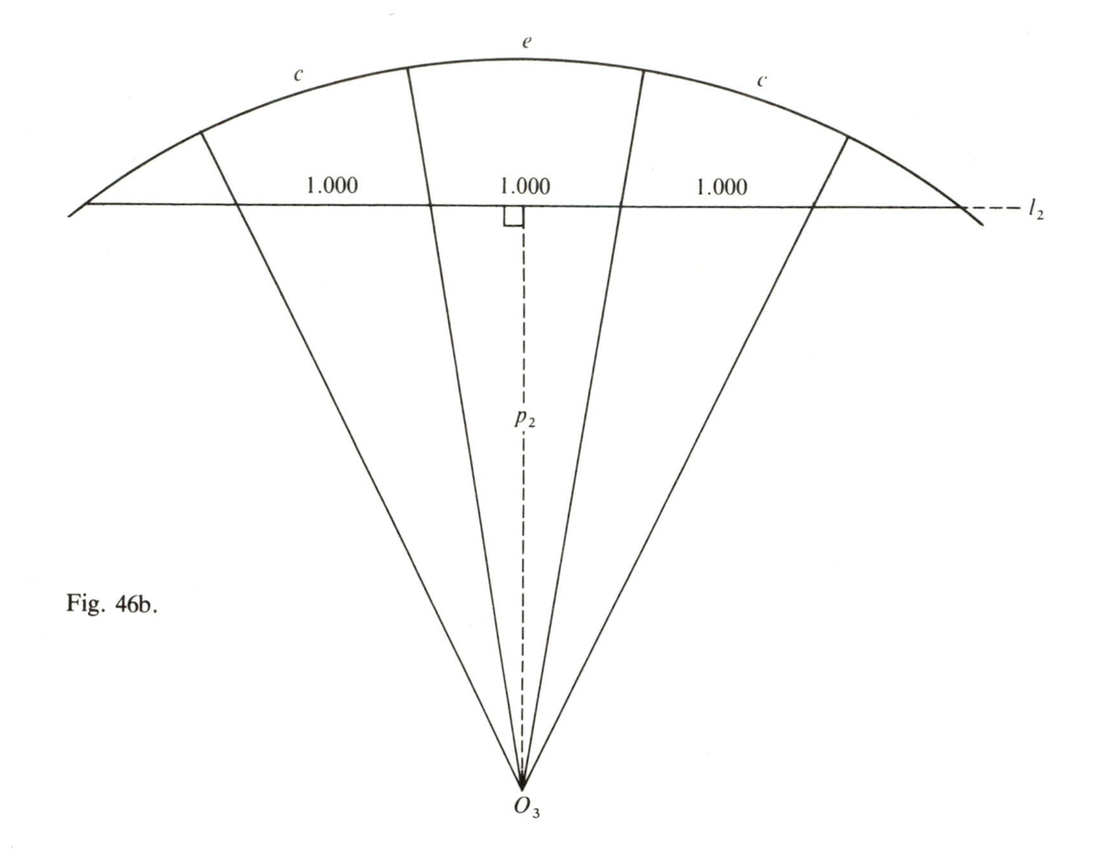

Fig. 46b.

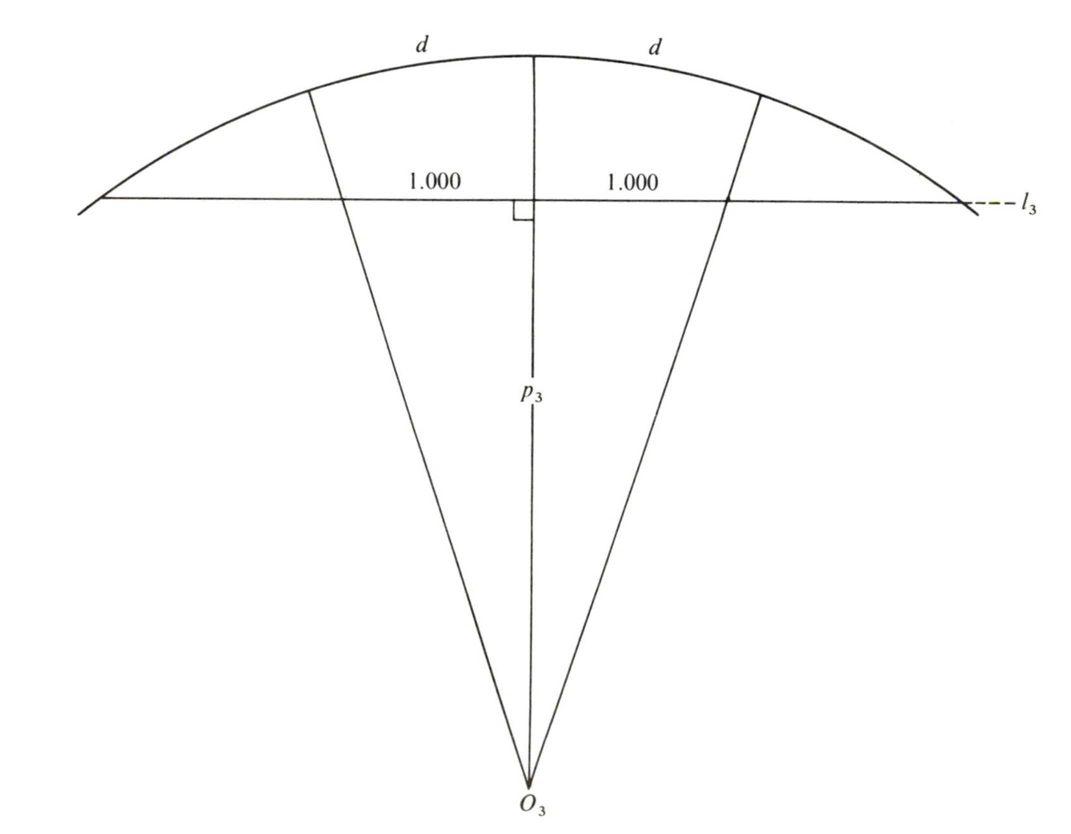

Fig. 46c.

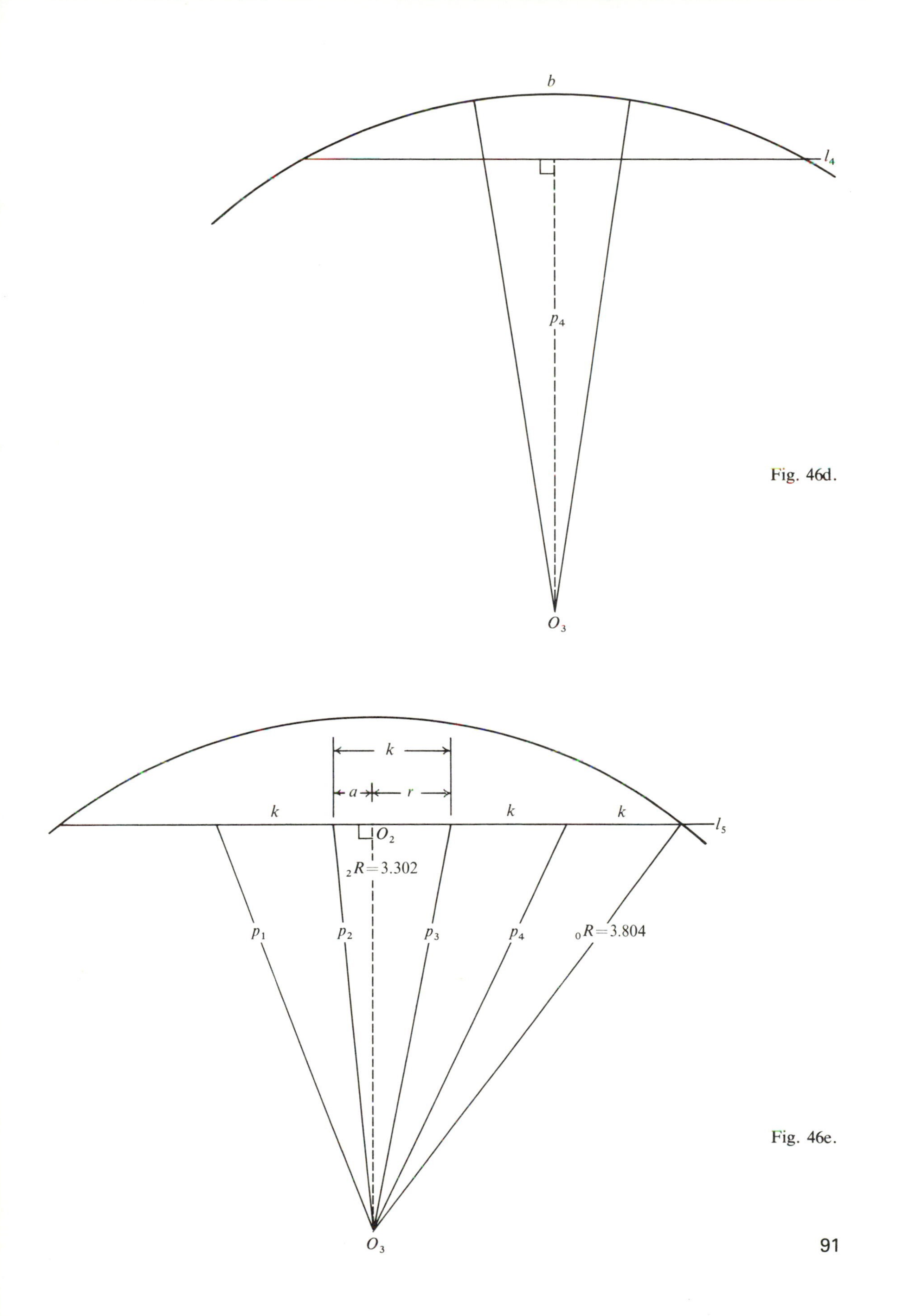

Fig. 46d.

Fig. 46e.

General instructions for making geodesic models

As the models now become more complex, careful attention is needed to get all the triangles in their proper places. If you write the letters a, b, c, d, e, f on the paper bands, this helps to identify them and you can glue the bands together by setting a to a, b to b, etc., so that the writing disappears between the bands. To achieve this, the writing must be done on opposite sides of the paper for right- and left-handed parts.

The assembly can be done by first making suitable subgroupings. This may be an entire icosahedral face, but this is not recommended for high-frequency domes. The best procedure, you may discover by experience, is to begin by making one pentagonal arrangement that uses five of **1**. When you see that this fits together well, add to it five of **2**. Then fill in the spaces between with ten of **3** in right- and left-handed pairs. This completes another and larger pentagonal arrangement. In this way you begin at what may be called the north pole of the dome and, by adding rows of bands or other subgroups, you work around the pole to approach the equator. Then continue in the same way around the equator till you reach the south pole. It is a real thrill to see the sphere develop in this way and to see it neatly close itself off with the last pentagonal subgroup. Tweezers and clamps or paper clips are useful construction tools.

It will be left for you to try out your own ingenuity on higher-frequency segmentations. Even if you do no more than just the drawings for some of these, you will notice that the arc length nearest the vertex O_0 is relatively short compared to other arc lengths. For example, in the 4-frequency dome this arc length $a = 14.545$, whereas $b = 16.978$. This situation only grows worse as the frequency grows larger. The only remedy is to change over to a different mode of segmentation.

An alternative method of approaching geodesic segmentation

Figure 47 shows a 4-frequency segmentation in which the chord l_1 does not have equal linear segments. But, as you can see in the related great circle arc shown in Fig. 47a, the arc lengths marked a have been chosen equal and these, in turn, determine the points of segmentation on l_1. This creates a difficulty, however, when the results are carried back to the flat icosahedron face. What may be called "windows" will begin to appear on the interior portion of the face. Even if the central point of the window is chosen as the point to be projected onto the surface of the sphere, you can well imagine that the continued use of a purely geometrical method of finding the required arc lengths would become a rather messy affair, to say nothing of the trigonometry involved in calculating the arc lengths.

The following elaboration is, therefore, given here as a simple method by which the dedicated model maker may skirt the problems that such windows introduce. It is a method that admittedly gives only approximate, but still very good, results for model making. This method, however, must use a formula from spherical trigonometry.

Books on spherical trigonometry contain many formulas, a fact that makes the average mathematics student shy away from the topic. But it is also a fact that all of these formulas are ultimately derived from plane geometry and trigonometry. When you see how these formulas can obtain good results very easily, you may be inspired to go on to study the entire topic of spherical trigonometry. (See especially

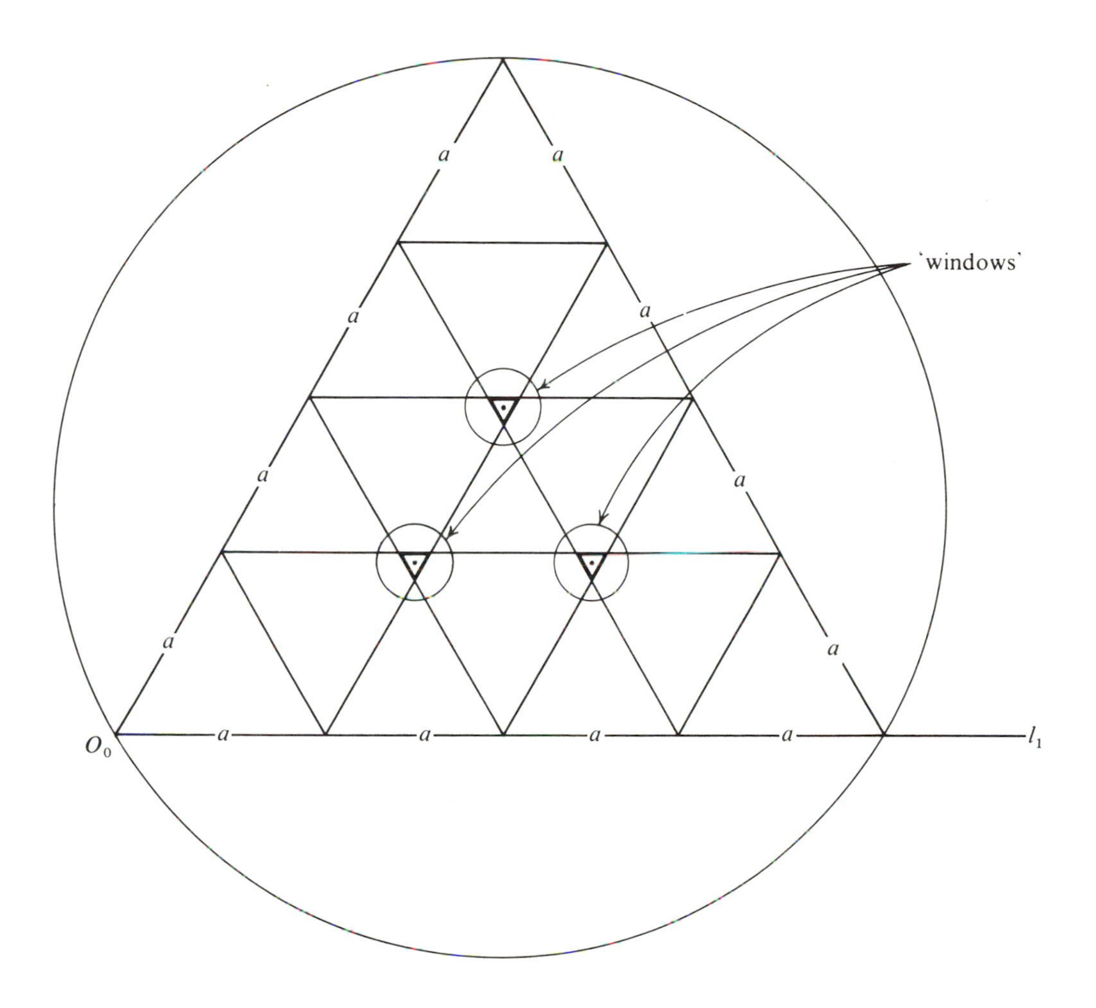

Fig. 47. A 4-frequency segmentation leading to "windows" in a triangle face of an icosahedron.

Chapter VIII of Hogben for good introductory material.)

Some general formulas from spherical trigonometry are given here for future use, should you want them or need them.

In any right spherical triangle (see Fig. 47b),

$$\sin a = \sin A \sin c$$
$$\tan b = \cos A \tan c$$
$$\cos c = \cos a \cos b$$

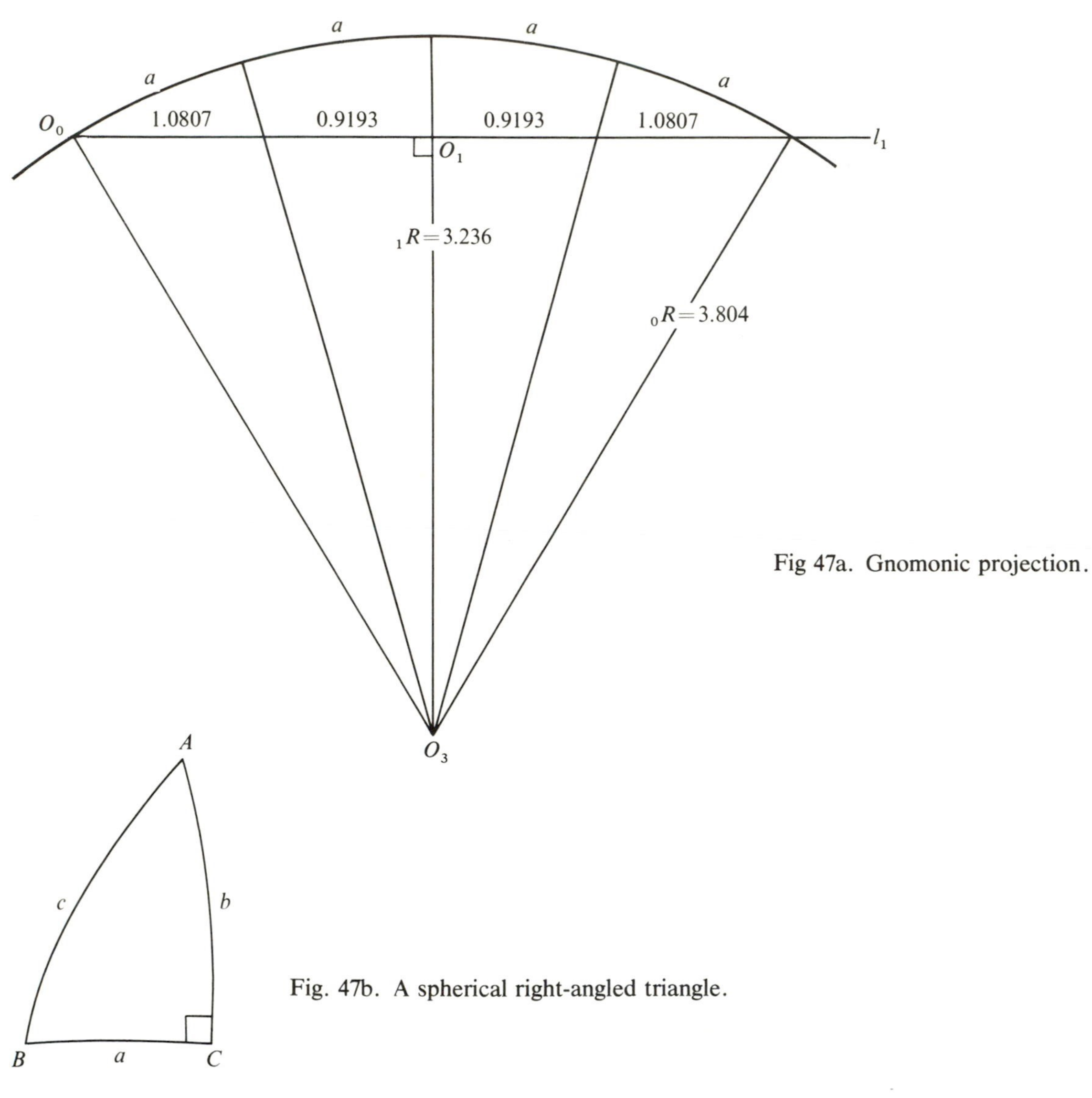

Fig 47a. Gnomonic projection.

Fig. 47b. A spherical right-angled triangle.

For any spherical triangle (see Fig. 47c),

$$\frac{\sin A}{\sin a} = \frac{\sin B}{\sin b} = \frac{\sin C}{\sin c}$$

$$\cos a = \cos b \cos c + \sin b \sin c \cos A$$

It is interesting to compare these formulas with those given for ordinary trigonometry.

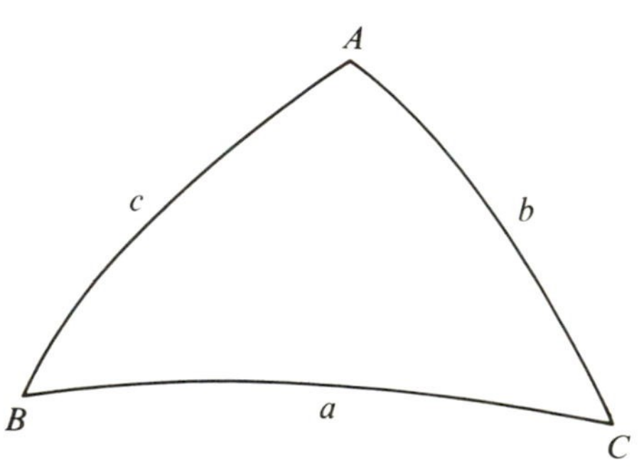

Fig. 47c. Any spherical triangle.

Fig. 48. A 4-frequency icosahedron.

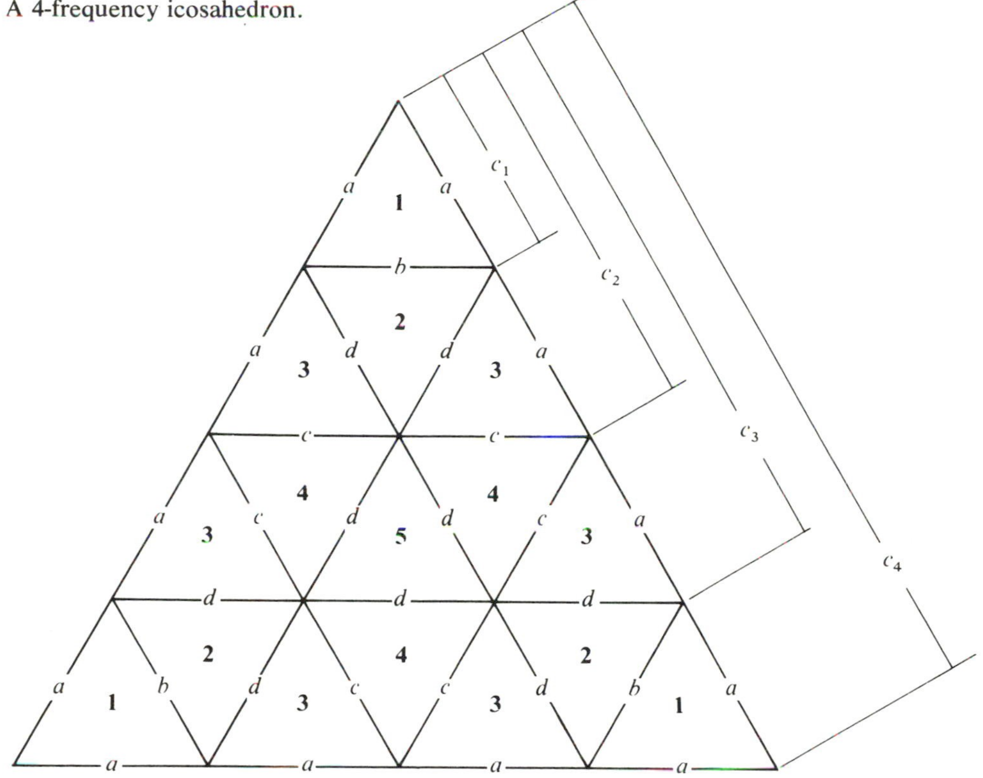

A 4-frequency model

The first formula will now be given extensive use. Consider a 4-frequency icosahedron as shown in Fig. 48. The flat triangle shown here must be given a spherical interpretation, in other words, as already projected onto the surface of a sphere. In the spherical formula given with Table 4, the subscript i is introduced to designate the number of segments of arc along an edge. Here c_i is the independent variable, a_i is the dependent variable, whereas A remains a constant. The measure of A is 36° since it is at the center of a pentagonal group. If you have a small hand-held electronic calculator available, you may perhaps enjoy verifying the results presented in Tables 4–6.

Table 4. *The 4-frequency model*

Sin a_i = sin A sin c_i				
i	1	2	3	4
c_i	15.859	31.717	47.576	63.453
a_i	9.243	18.000	25.715	31.717

Letters a and c of Fig. 48 must not be confused with a_i and c_i of Table 4. Their relationship is shown in the arc measures given here for making this model:

arcs	*bands*
$a = c_1 = 15.859$	**1** *a b a*
$b = 2a_1 = 18.486$	**2** *d b d*
$c = a_2 = 18.000$	**3** *a c d*
$d = \frac{2}{3}a_3 = 17.143$	**4** *c d c*
	5 *d d d*

6-Frequency and 8-frequency models

The data for a 6-frequency and an 8-frequency dome, are listed in Tables 5 and 6, respectively.

Table 5. *The 6-frequency model*

Sin a_i = sin A sin c_i						
i	1	2	3	4	5	6
c_i	10.572	21.145	31.717	42.290	52.862	63.435
a_i	6.191	12.241	18.000	23.298	27.942	31.717

arcs

$a = c_1 = 10.572$

$b = \frac{2}{1} a_1 = 12.382$

$c = \frac{2}{2} a_2 = 12.241$

$d = \frac{2}{3} a_3 = 12.000$

$e = \frac{2}{4} a_4 = 11.649$

$f = \frac{2}{5} a_5 = 11.177$

$\frac{2}{6} a_6 = 10.572 = a$

bands

1 *a b a*
2 *f b f*
3 *a c f*
4 *f d f*
5 *c e f*
6 *a d e*
7 *e d e*
8 *e f e*
9 *d f d*

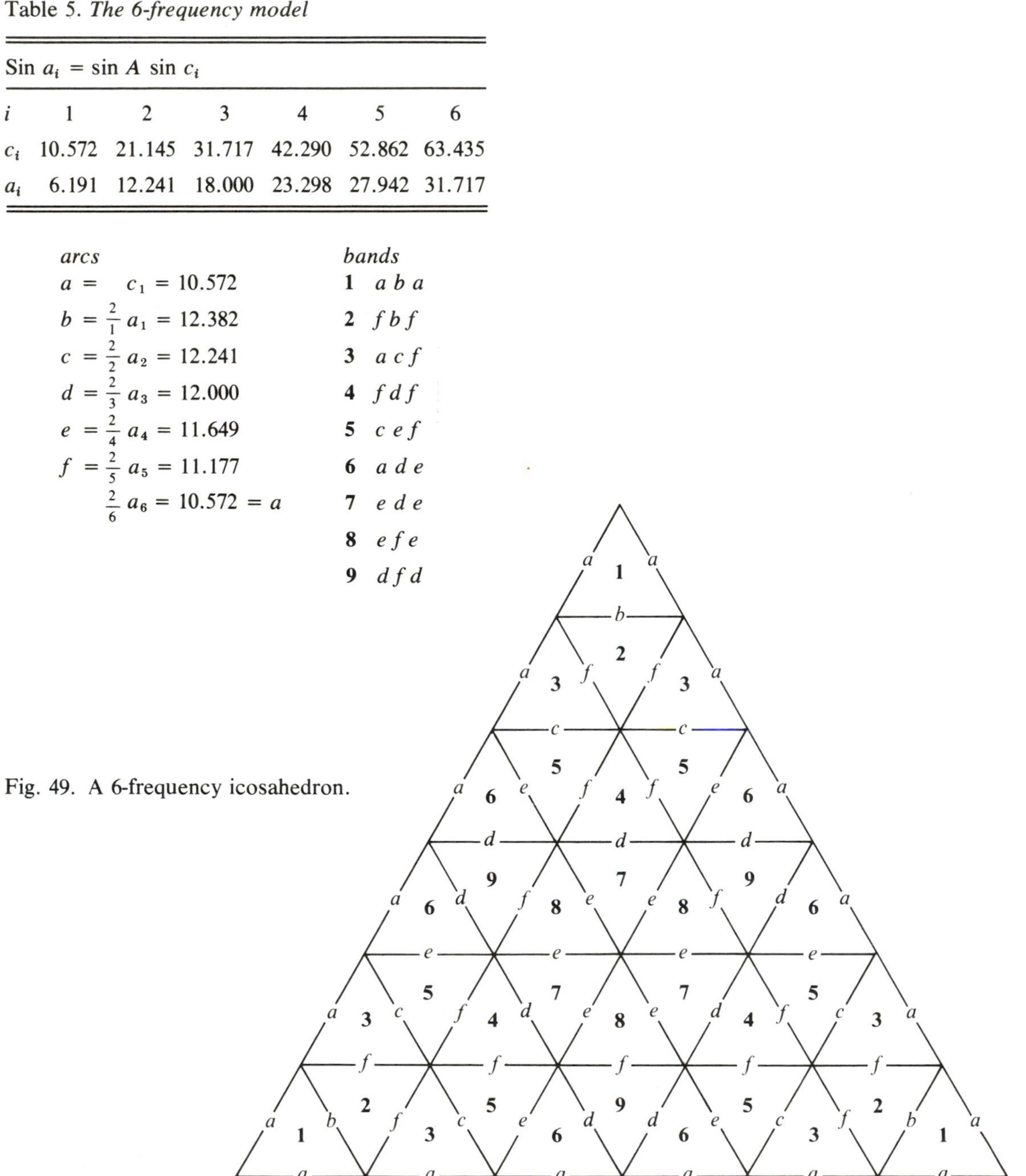

Fig. 49. A 6-frequency icosahedron.

It is interesting to see a serial relationship developing here, present already in the lower frequencies but now becoming more pronounced.

Table 6. *The 8-frequency model*

Sin a_i = sin A sin c_i								
i	1	2	3	4	5	6	7	8
c_i	7.929	15.859	23.788	31.717	39.647	47.576	55.506	63.435
a_i	4.651	9.243	13.715	18.000	22.027	25.714	28.976	31.717

arcs

$a = c_1 = 7.929$

$b = \frac{2}{1} a_1 = 9.302$

$c = \frac{2}{2} a_2 = 9.243$

$d = \frac{2}{3} a_3 = 9.143$

$e = \frac{2}{4} a_4 = 9.000$

$f = \frac{2}{5} a_5 = 8.811$

$g = \frac{2}{6} a_6 = 8.571$

$h = \frac{2}{7} a_7 = 8.279$

$\frac{2}{8} a_8 = 7.929 = a$

bands

1 *a b a*
2 *h b h*
3 *a c h*
4 *h d h*
5 *c g h*
6 *a d g*
7 *g d g*
8 *e g h*
9 *d f h*
10 *a e f*
11 *g f g*
12 *e f g*
13 *f h f*
14 *e h e*
15 *f f f*

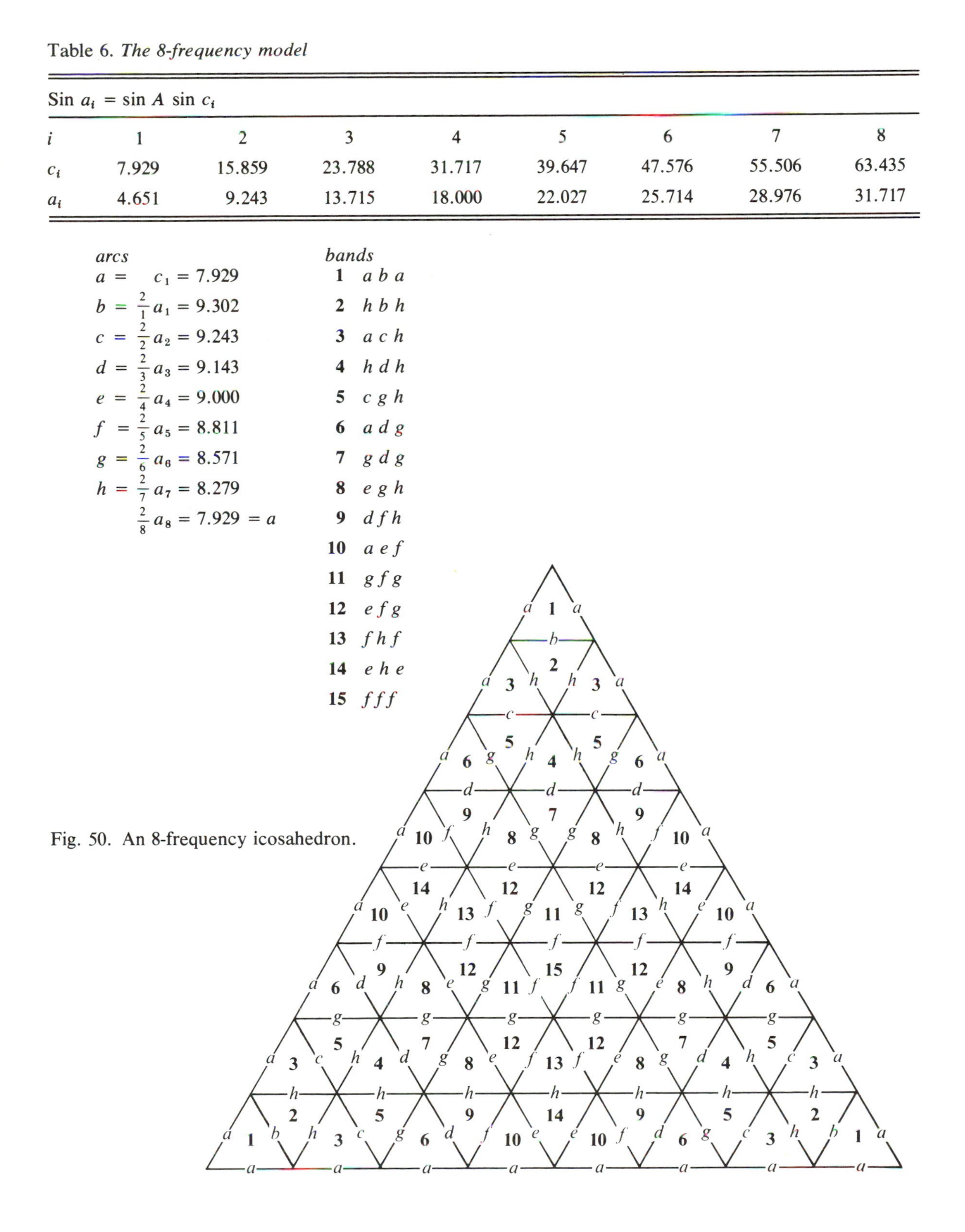

Fig. 50. An 8-frequency icosahedron.

Photo 39. $\{3, 5+\}_{6,0}$

Photo 40. $\{3, 5+\}_{8,0}$

It is truly amazing what a small calculator can do. It can give you results that are often correct to eight significant figures. For model making, you need at most an arc length correct to the nearest tenth of a degree, so calculations are best done by retaining at least three places of decimals. Even so you mark your protractor as best you can at points between the half-degree marks. This gives sufficient accuracy at the scale used here for domes up to a diameter of 12 in. or about 30 cm.

Although the results given in Tables 4–6 are correct so far as the formula is concerned, the use of these results is strictly speaking incorrect. The "windows" do not really disappear entirely, but in moving from the plane face of the icosahedron to its spherical counterpart, the window effect becomes even less pronounced. Finding all the arc lengths after segmenting the edges into equal arc lengths would require the full range of formulas from spherical trigonometry. The slight differences are negligible because, in making the models with paper bands as suggested here, these differences tend to disappear into the vertex areas. Photo 39 shows a 6-frequency dome and Photo 40 an 8-frequency dome done very successfully by using the results given in Tables 5 and 6. They turn out to be very attractive as well – see the frontispiece for the 8-frequency model in color.

Introduction to geodesic symbolism and classification

Before going ahead with higher-frequency geodesic domes based on polyhedrons other than the icosahedron, it may be good to introduce some symbolism by which to classify them and to count the number of vertices, faces, and edges. A geodesic dome is basically a polyhedron, so the Euler formula relating the number of vertices, faces, and edges applies: $V + F = E + 2$.

The following elaboration owes its inspiration to Professor Coxeter (see Coxeter, "Virus macromolecules and geodesic domes"). The symbol $\{3, 5+\}_{b,c}$ is used for a spherical tessellation in which all the faces are triangles and in which the number of faces $F = 20T$, the number of edges $E = 30T$, and hence, by the Euler formula the number of vertices, $V = 10T + 2$. In all three of these formulas $T = b^2 + bc + c^2$, where b and c are step counters, to be explained later. Notice that b and c are the subscripts in the symbol $\{3, 5+\}_{b,c}$. Your first reaction may be: Where do all these formulas come from? Mathematicians have a way of giving formulas or symbols that often leave the general reader amazed at the ingenuity with which they are devised. Often enough they give very meager indications as to how they themselves arrived at such insights. But no matter; the results may be satisfactorily used before complete understanding is attained. In fact seeing the formulas work may spur you on to study why they work. This may then be an incentive for going on to the study of higher mathematics, another name for more abstract mathematics, which is what symbols really are.

The symbol $\{3, 5+\}_{b,c}$ can best be understood by looking at models. Look first at the spherical icosahedron itself, a 1-frequency dome. All its 20 faces are triangles, the number of its edges is 30, and the number of its vertices is 12. Substitution in the formula $F = 20T$ and into the formula $E = 30T$ gives $T = 1$. In the formula $V = 10T + 2$ if $V = 12$, again $T = 1$. The model obviously has 12 pentavalent vertices. Since the total number of vertices is given by the formula $V = 10T + 2$, the number of hexavalent vertices can be found by subtracting 12 from this total. Thus the number of hexavalent vertices can be expressed as $10(T - 1)$. In this expression if $T = 1$, it tells us that the number of hexavalent vertices in this case is zero. You can see this is true.

Try all these formulas now on the spherical dodecahedron, the pentakisdodecahedron. Here $F = 60$, $V = 32$, making $E = 90$ and $T = 3$. The number of pentavalent vertices is 12 and the rest are hexavalent. If $T = 3$, then $10\ (T - 1) = 20$. You can see this is true.

The question still remains: How can b and c be evaluated? In the formula $T = b^2 + bc + c^2$ the values of b and c can easily be determined by looking at models. The procedure is as follows: With a model in your hands, first locate two neighboring pentavalent vertices and count the number of steps from one to the other along geodesic arcs, either b steps straight along a great circle, so that $c = 0$, or $b + c$ steps along a bent path, taking first b steps straight along a great circle, then turning either right or left and taking c steps to arrive at the neighboring pentavalent vertex.

With these ideas in mind, examine all the geodesic models you have made that were based on the icosahedron face or in the icosahedral symmetry group. If c remains zero, b is the frequency. If you examine the pentakisdodecahedron model, you can verify that $b = 1$ and $c = 1$. These values tie into the formula $T = b^2 + bc + c^2$. For the spherical icosahedron, $b = 1$ and $c = 0$, giving $T = 1$. For the spherical pentakisdodecahedron, $b = 1$ and $c = 1$, giving $T = 3$: For higher-frequency icosahedral domes, where c is always zero, $T = b^2$ alone; this means that the number of triangle elements is equal to the square of the frequency.

Knowing T you can now easily evaluate F, V, E by using the appropriate formulas. To see how this applies, take a look at the spherical snub dodecahedron as shown in Fig. 43 and Photo 36, assuming you have made the model. You will find in it that $b = 2$ and $c = 1$. Its symbol therefore is $\{3, 5+\}_{2,1}$. Here $T = 4 + 2 + 1 = 7$, from which it follows that the number of faces $F = 140$, the number of edges $E = 210$, and the number of vertices $V = 72$, of which 12 are pentavalent and the remaining 60 are hexavalent. Notice that the count has gone up fast, but now for any desired value of b and c there is no great difficulty either in classifying the type of geodesic dome it is, nor in arriving at the number of its faces, edges, and vertices. It should be noted here that $b + c$ always equals the frequency. Thus the spherical snub dodecahedron is really a 3-frequency geodesic dome.

The spherical truncated icosahedron is now classified as $\{3, 5+\}_{3,0}$, and the spherical pentakisdodecahedron is $\{3, 5+\}_{1,1}$.

Photo 41. $\{3, 5+\}_{2,2}$

Geodesic models derived from the dodecahedron

It will now be shown how to obtain a model of $\{3, 5+\}_{2,2}$ (see Photo 41). The regular dodecahedron may be used for this. One of the five isosceles triangles on a pentagon face of the dodecahedron can be given the same segmentation treatment as was given to the equilateral triangle face of the icosahedron. Figure 51 suggests the way this is done. Arcs a, d, e are easily calculated from Figs. 51a and 51d; arc b can be found by first determining p_2 in Fig. 51b. Arc c is a bit more

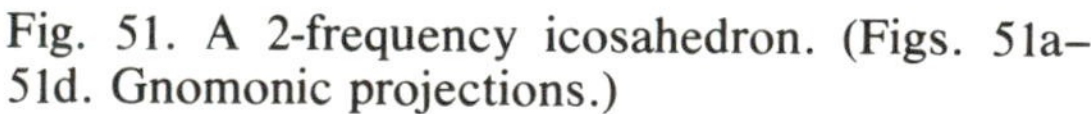

Fig. 51. A 2-frequency icosahedron. (Figs. 51a–51d. Gnomonic projections.)

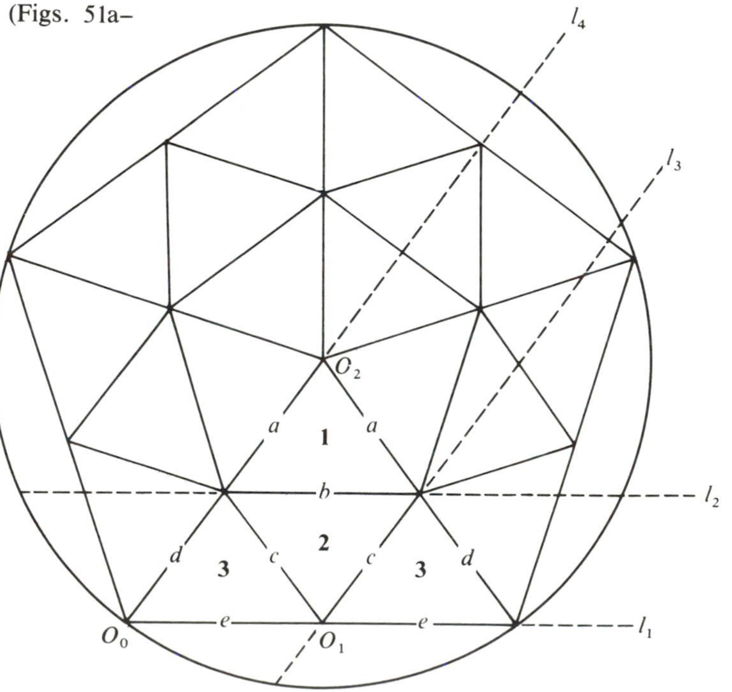

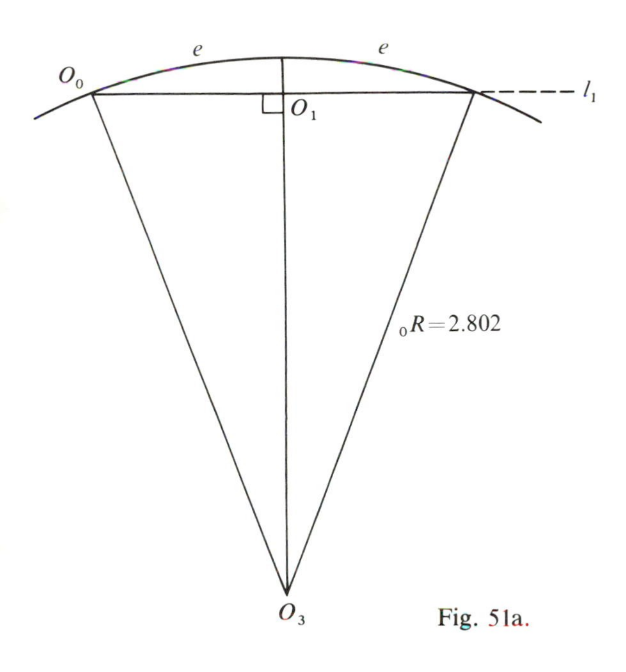

Fig. 51a.

troublesome, but it can be found from the scalene triangle that appears in Fig. 51c. On a small-scale model, no great harm can come from relying completely on the accuracy of your geometrical constructions alone. But the formulas from spherical trigonometry can be applied very simply here, if you want to check out the numerical data. The arc measures and bands are as follows:

arcs	*bands*
$a = 20.905$	**1** *a b a*
$b = 24.214$	**2** *c b c*
$c = 18.841$	**3** *c e d*
$d = 16.472$	
$e = 20.905$	

The process by which higher-frequency domes are obtained with $b = c$ should now be clear. You need only increase the number of segments in the isosceles triangle on the face of the dodecahedron. But the gnomonic projection becomes increasingly complex, and the results are not good because the arc lengths near a vertex O_0 tend to decrease rapidly when

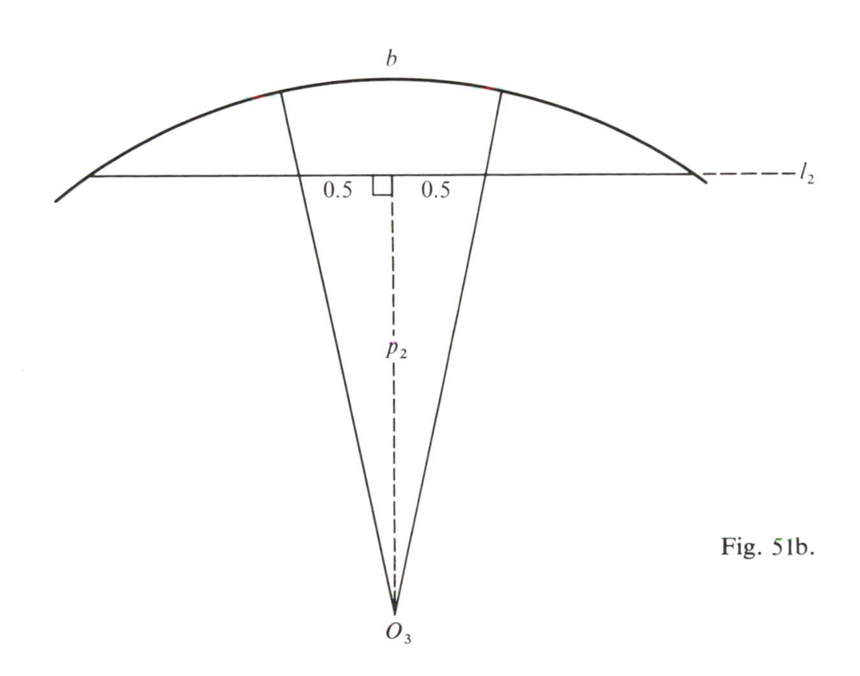

Fig. 51b.

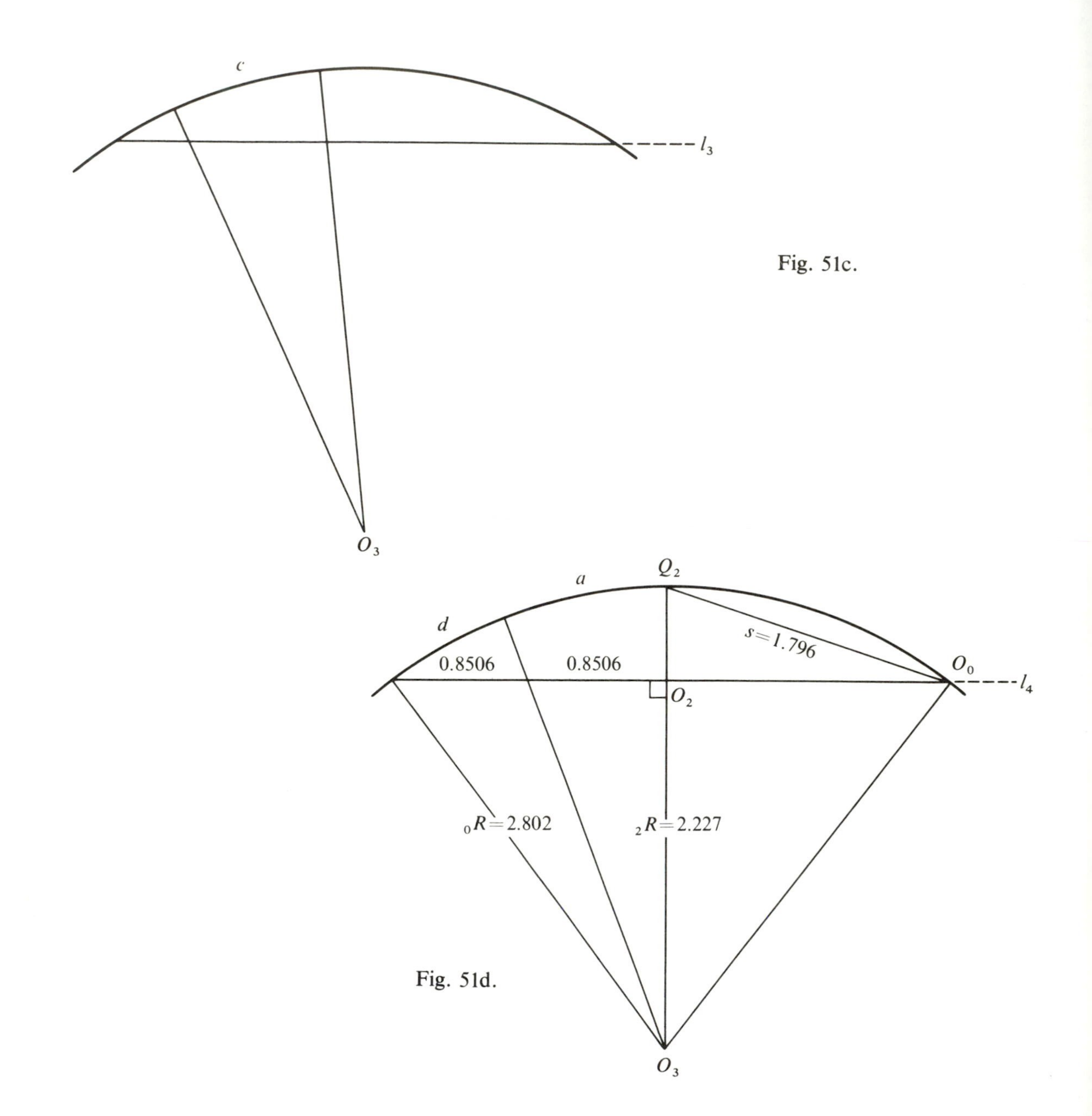

Fig. 51c.

Fig. 51d.

the frequency becomes greater than 2. Therefore it is better to shift to a pentakisdodecahedron and apply the segmentation process to one of its sixty, flat, isosceles triangle faces – Fig. 52 shows how to begin again with a 2-frequency segmentation. The base line or chord l_1 is still equivalent to the dodecahedron edge, but the lateral sides of this isosceles triangle have the measure $s = 1.796$ when $e = 2$. The length s is geometrically displayed in Fig. 51d. You will recognize here that the point O_2, the incenter of the dodecahedron face, has been projected to the point Q_2, making $O_0Q_2 = s$. The method of gnomonic projection to obtain arc lengths for a

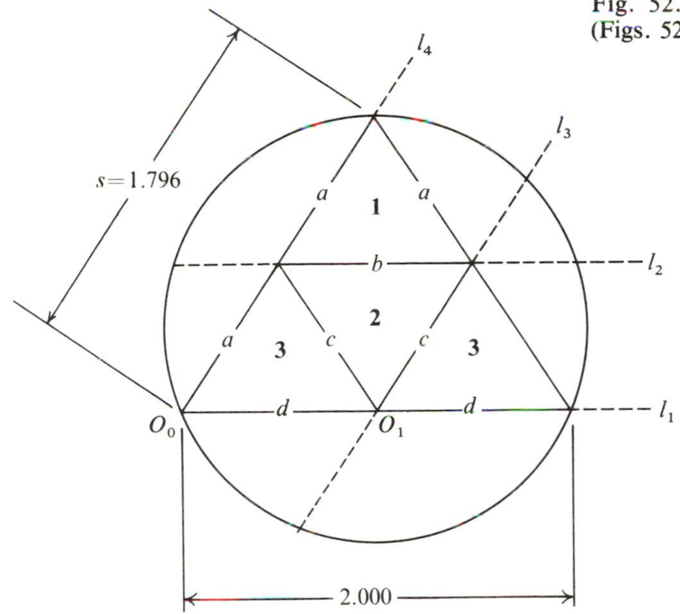

Fig. 52. A 2-frequency pentakisdodecahedron. (Figs. 52a–52d. Gnomonic projections.)

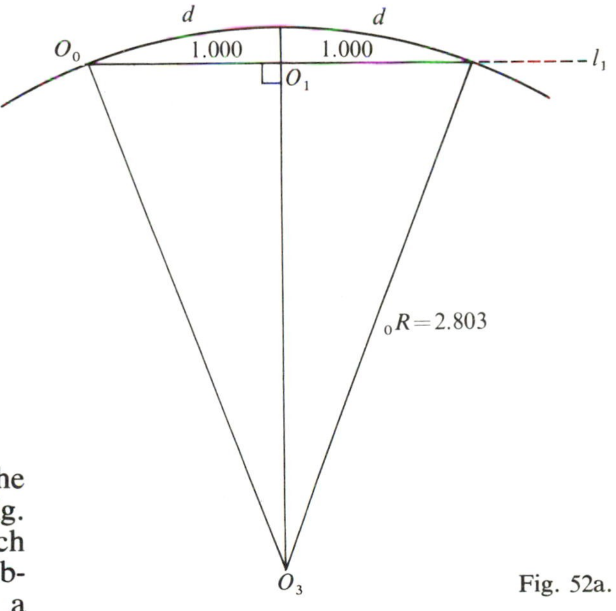

Fig. 52a.

model of $\{3, 5+\}_{2,2}$ now proceeds in the usual way. The isosceles triangle of Fig. 52 is given its circumscribing circle, which is a small circle of the sphere circumscribing the basic polyhedron, in this case a pentakisdodecahedron. Figures 52a–52d all show great circle arcs of radius $_0R$ = 2.803, because the basic polyhedron here

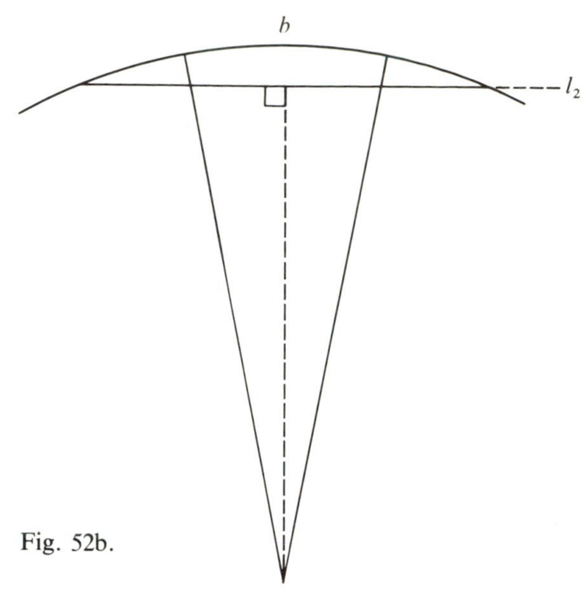

Fig. 52b.

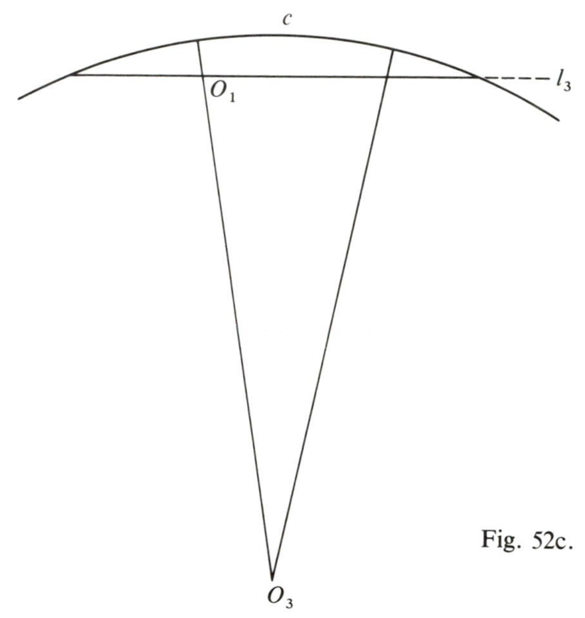

Fig. 52c.

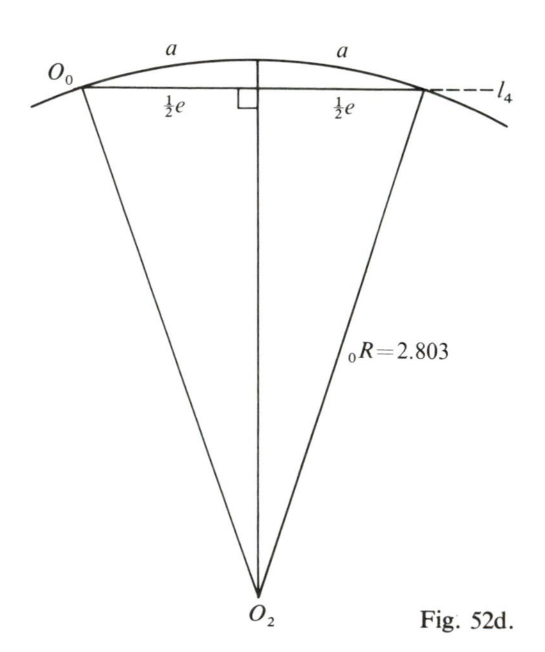

Fig. 52d.

is really still a spherical version of the dodecahedron. The arc measures and bands are as follows:

arcs	*bands*
$a = 18.689$	**1** $a\ b\ a$
$b = 21.712$	**2** $c\ b\ c$
$c = 19.595$	**3** $a\ d\ c$
$d = 20.905$	

Figure 53 shows the tessellation for a model of $\{3, 5+\}_{3,3}$, and Figs. 53a–53d show the gnomonic projections. Figure 53e illustrates the method by which p_1, p_2, p_3 can be calculated and from which the

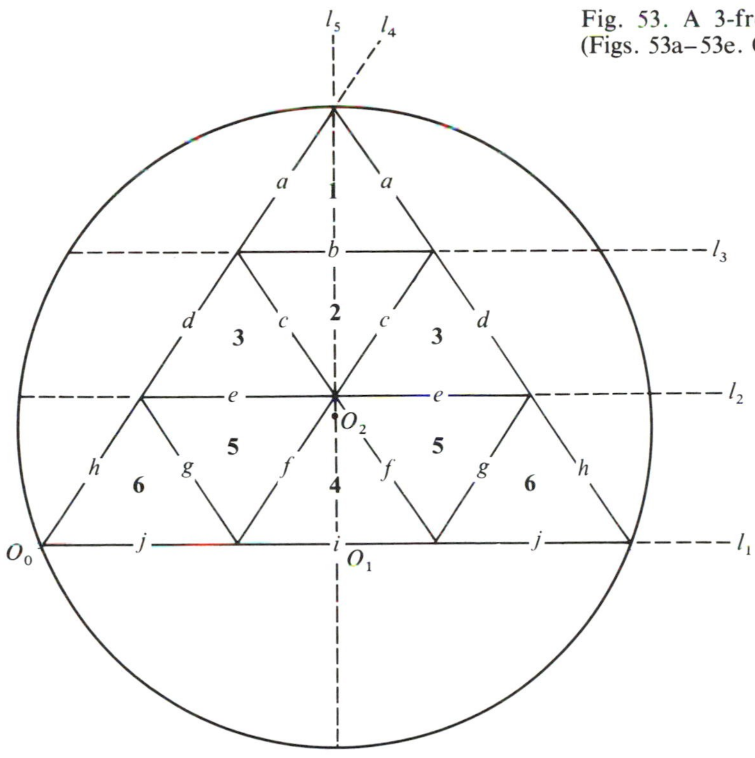

Fig. 53. A 3-frequency pentakisdodecahedron. (Figs. 53a–53e. Gnomonic projections.)

Photo 42. $\{3, 5+\}_{3,3}$

arc measures are found. But you may by now have learned that the formulas from spherical trigonometry are more easily applied. For this 3-frequency model, the arc measures and bands are as follows:

arcs	*bands*
a = 12.256	**1** *a b a*
b = 14.335	**2** *c b c*
c = 12.942	**3** *c e d*
d = 12.865	**4** *f i f*
e = 14.450	**5** *f e g*
f = 13.106	**6** *g j h*
g = 12.926	
h = 12.256	
i = 14.512	
j = 13.649	

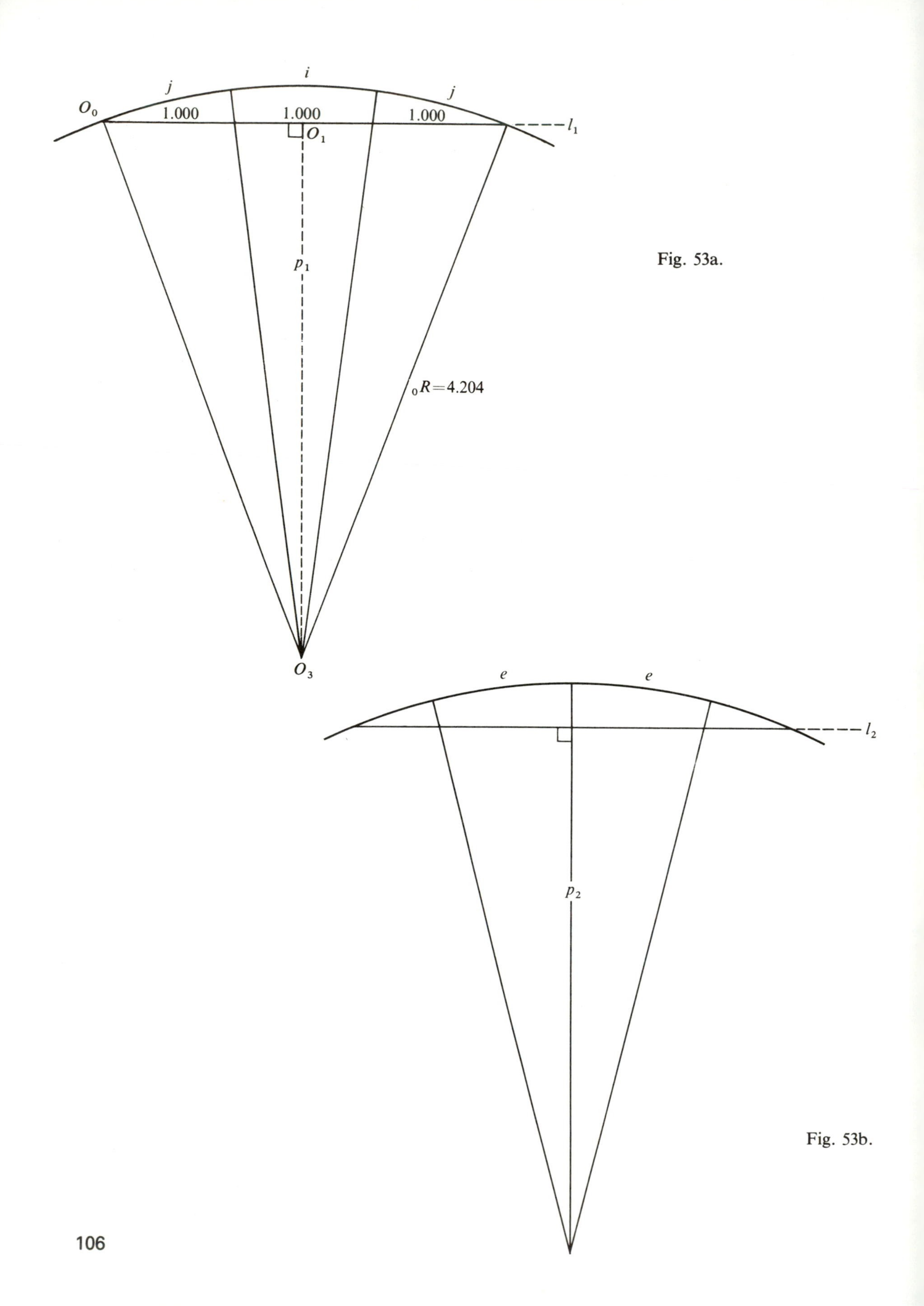

Fig. 53a.

Fig. 53b.

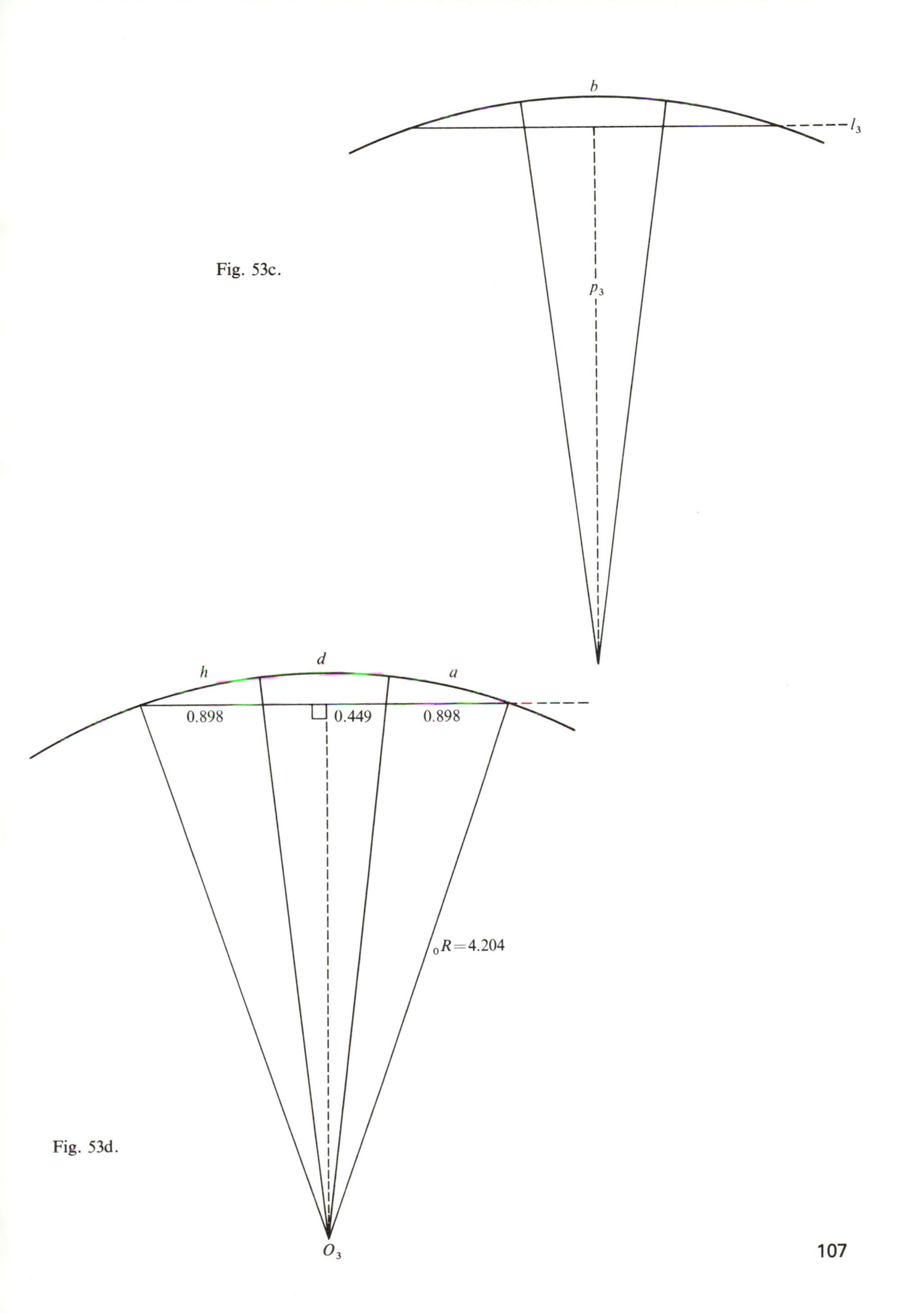

Fig. 53c.

Fig. 53d.

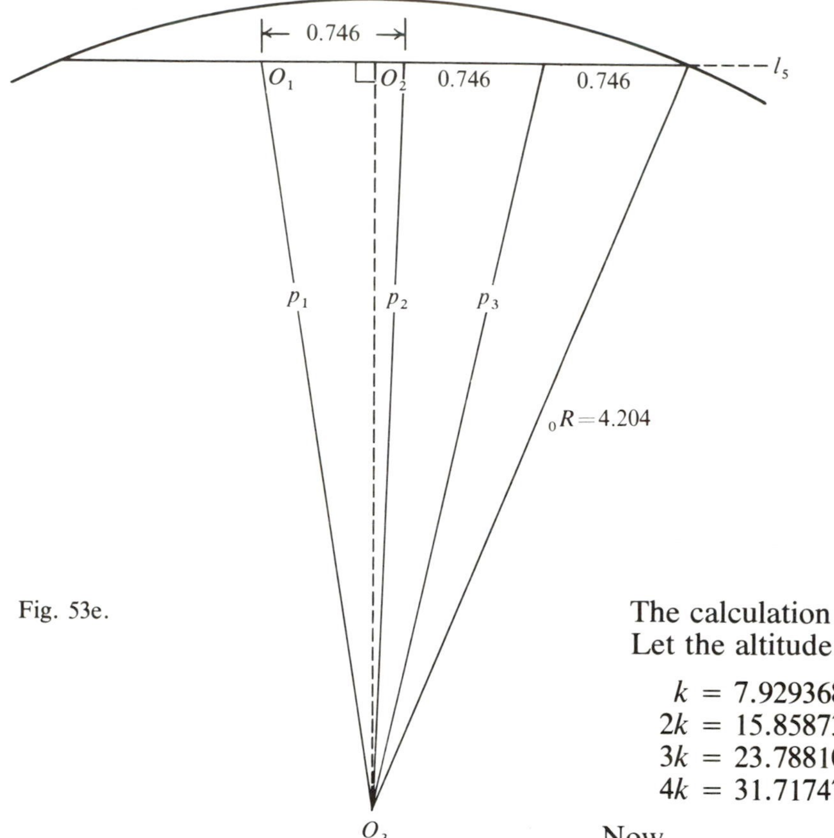

Fig. 53e.

The calculation is done as follows: Let the altitude equal $4k$. Then

$$k = 7.9293685 = (1/4)(31.717474)$$
$$2k = 15.858737$$
$$3k = 23.788106$$
$$4k = 31.717474$$

Now

$$\tan a = \frac{\tan k}{\cos 36}, \quad \text{from which } a = 9.7685315$$

$$\tan(a + c) = \frac{\tan 2k}{\cos 36}, \quad \text{from which } c = 9.5797345$$

$$\tan(a + c + e) = \frac{\tan 3k}{\cos 36}, \quad \text{from which } e = 9.236130$$

$$\tan(a + c + e + g) = \frac{\tan 4k}{\cos 36}, \quad \text{from which } g = 8.792972$$

Finally

$$\sin \tfrac{1}{2}b = \sin a \sin 36, \quad \text{from which } b = 11.447073$$
$$\sin d = \sin(a + c) \sin 36, \quad \text{from which } d = 11.229462$$
$$\sin(\tfrac{1}{2}b + f) = \sin(a + c + e) \sin 36, \quad \text{from which } f = 10.609945$$
$$\sin(h + d) = \sin(a + c + e + g) \sin 36, \quad \text{from which } h = 9.6756984$$

An alternative for geodesic segmentation of the dodecahedron

For frequencies higher than 3, segmenting the lateral edge of a pentakisdodecahedron is not a good method to use, because you see how the number of arc lengths grows rapidly. So a simpler method is that of segmenting the geodesic arc that forms the altitude of the isosceles triangle shown in Fig. 52 or Fig. 53. This method as it pertains to a model of $\{3, 5+\}_{4,4}$ is shown in Fig. 54. You see here that symmetry greatly reduces the number of arc lengths needed. Furthermore, the calculations are far more easily done by the formulas of spherical trigonometry than from projection drawings.

Fig. 54. A 4-frequency pentakisdodecahedron.

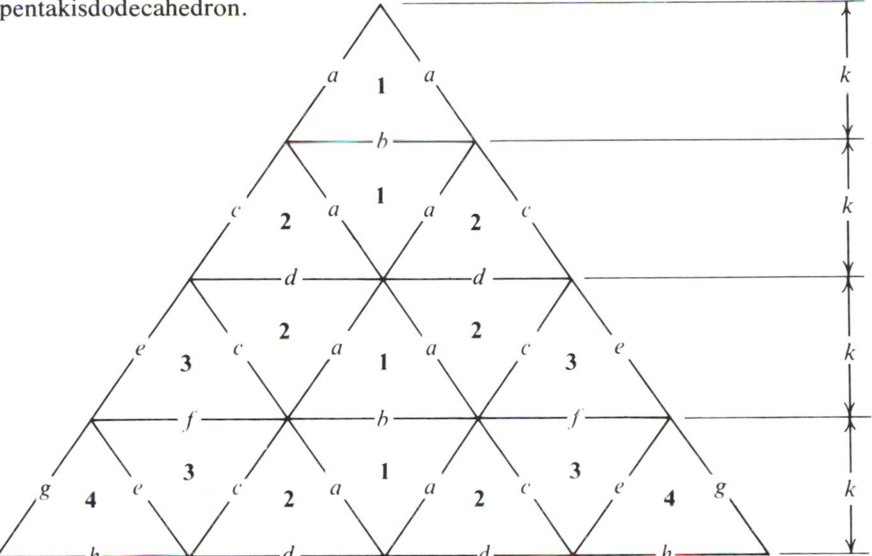

Thus,

arcs	*bands*
$a =$ 9.769	**1** $a\ b\ a$
$b =$ 11.447	**2** $a\ c\ d$
$c =$ 9.580	**3** $c\ e\ f$
$d =$ 11.229	**4** $e\ g\ h$
$e =$ 9.236	
$f =$ 10.610	
$g =$ 8.793	
$h =$ 9.676	

Note that here

$$(a + c + e + g) = 37.377368$$
$$4k = 31.717474$$
$$(h + d) = 20.905158$$

These are precisely the arc lengths belonging to the characteristic triangle of the icosahedral symmetry group. This triangle is thus a sort of "court of last appeal" for all geodesic segmentation in the icosahedral symmetry group. These values can also be found on a small calculator by the key punches:

$$(\tfrac{1}{2}) \arcsin \tfrac{2}{3} = a = 20.905158$$
$$(\tfrac{1}{2}) \arctan 2 = b = 31.717474$$
$$90 - a - b = c = 37.377368$$

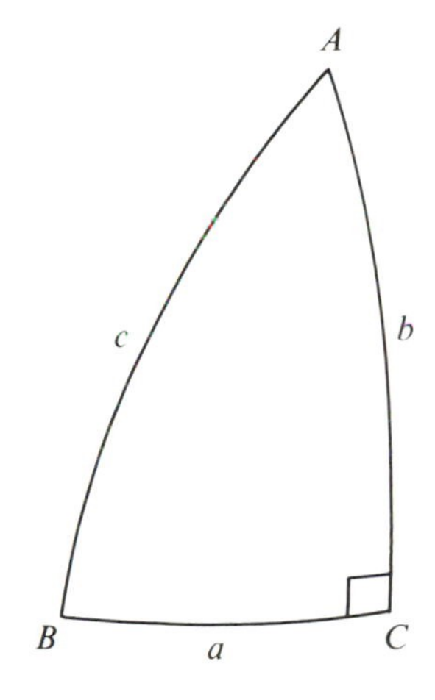

Fig. 54a. The characteristic triangle of the icosahedral symmetry group.

(see Fig. 54a). The dihedral angles A, B, C are of course the P, Q, R of Fig. 16, so that

$$P = A = 36$$
$$Q = B = 60$$
$$R = C = 90$$

A second alternative for geodesic segmentation of the icosahedron

A further development may now be presented. It is possible to break the icosahedral face into other tessellated networks of smaller triangles and gnomonically project chord segments to obtain arc lengths needed for making other models. For example, if perpendiculars are drawn from points of equal linear subdivision along icosahedral edges, a different type of tessellated network of smaller triangles is obtained. (See Fig. 55 for an example of $\{3, 5+\}_{2,2}$; Figs. 55a and 55b show the gnomonic projections.)

For this model, the arc measures and bands are as follows:

arcs	*bands*
$a = 16.472$	**1** *a b a*
$b = 19.188$	**2** *c b c*
$c = 20.554$	**3** *d c d*
$d = 20.905$	

Although the icosahedral face is used here, the assembly of the model must be done in the symmetry of the dodecahedron, because icosahedral edges do not appear as bands. In fact some of the bands straddle the icosahedral edge, as you can see by examining the drawings. Strictly speaking, arc *b* is not identically the same as twice the arc named (½)*b*, shown in Fig.

Fig. 55. A 4-frequency Class II icosahedron. $\{3, 5+\}_{2,2}$. (Figs. 55a, 55b. Gnomonic projections.)

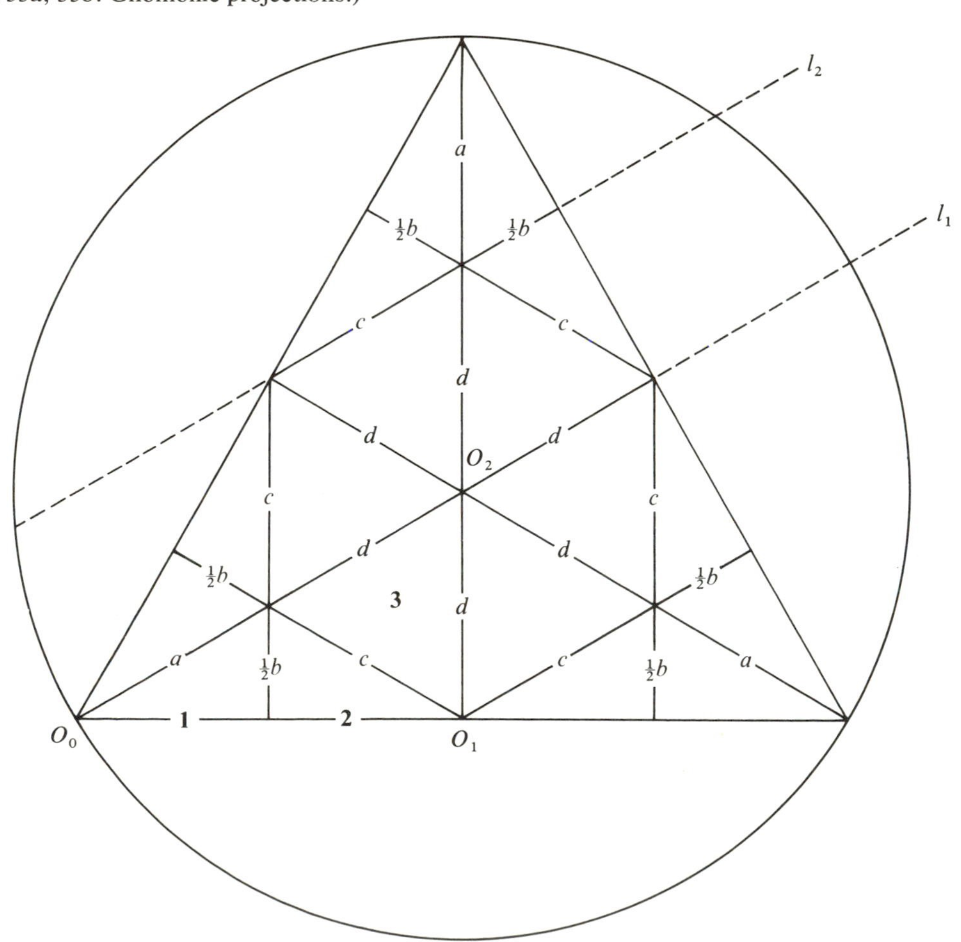

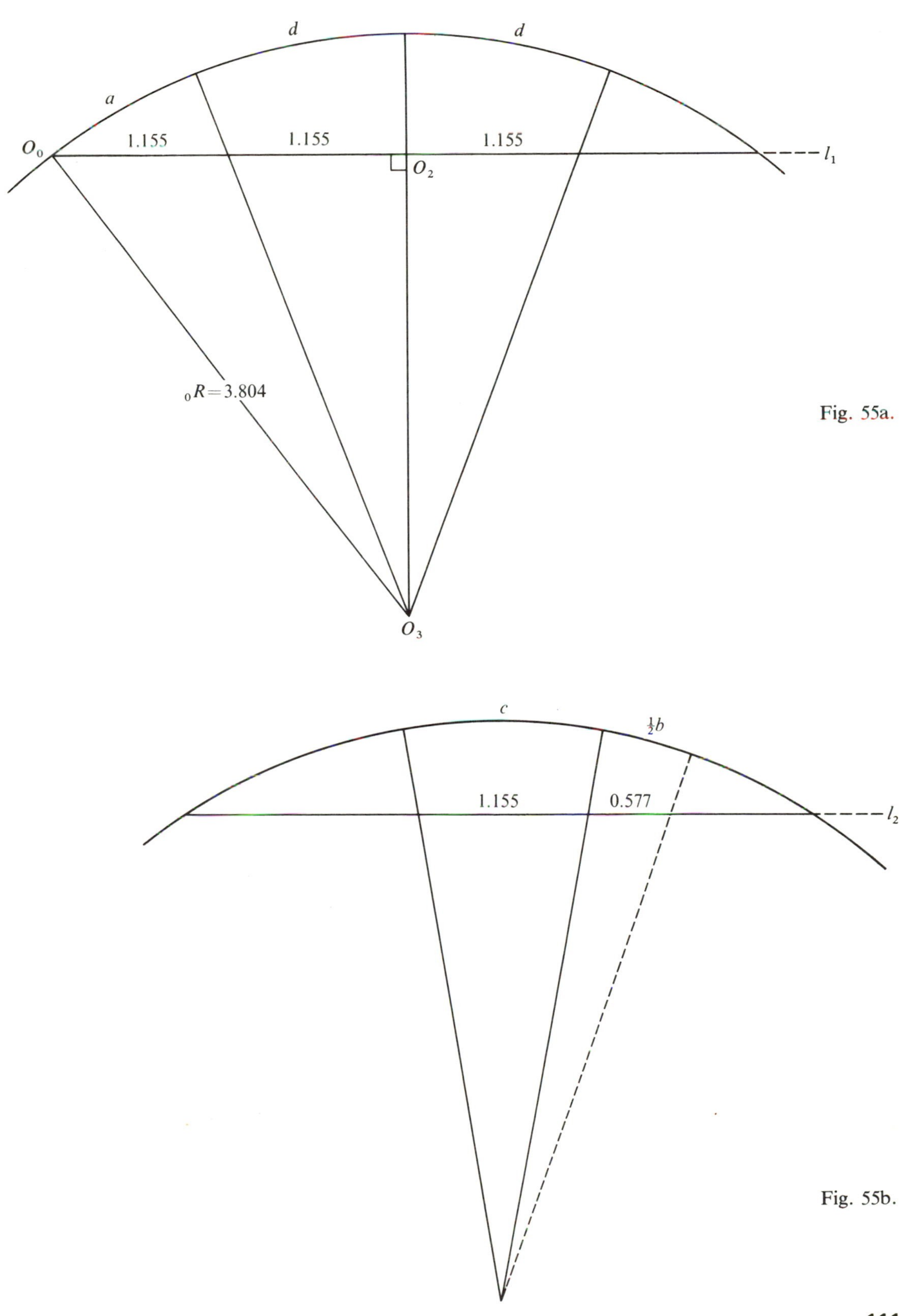

Fig. 55a.

Fig. 55b.

55b; but for model making the difference is negligible. However, the exact calculations can be done by using the formulas of spherical trigonometry.

Figure 56 shows the icosahedral face for $\{3, 5+\}_{3,3}$; Figs. 56a–56c are the gnomonic projections. The comment made above about arc *b* applies here to both arc *b* and arc *h*.

For this model, the arc measures and bands are as follows:

arcs	*bands*
$a = 10.388$	**1** *a b a*
$b = 12.168$	**2** *c b c*
$c = 13.115$	**3** *c e d*
$d = 12.703$	**4** *e h e*
$e = 13.368$	**5** *e f e*
$f = 14.175$	**6** *g f g*
$g = 14.286$	
$h = 13.238$	

Note that the same symbol, for example $\{3, 5+\}_{3,3}$, can belong to geodesic domes with different sets of arc lengths. This means that in general the symbol is, mathematically speaking, a topological one.

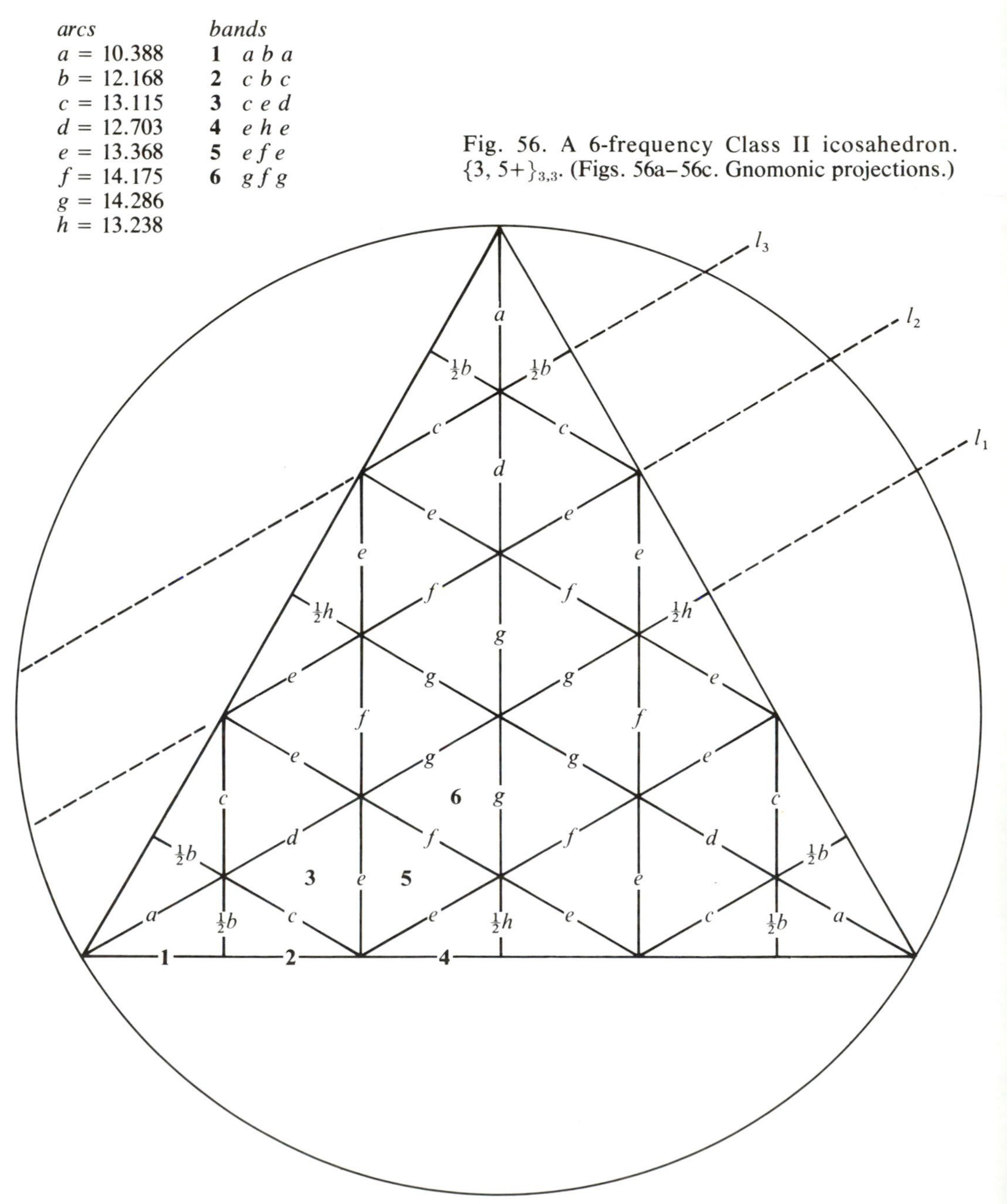

Fig. 56. A 6-frequency Class II icosahedron. $\{3, 5+\}_{3,3}$. (Figs. 56a–56c. Gnomonic projections.)

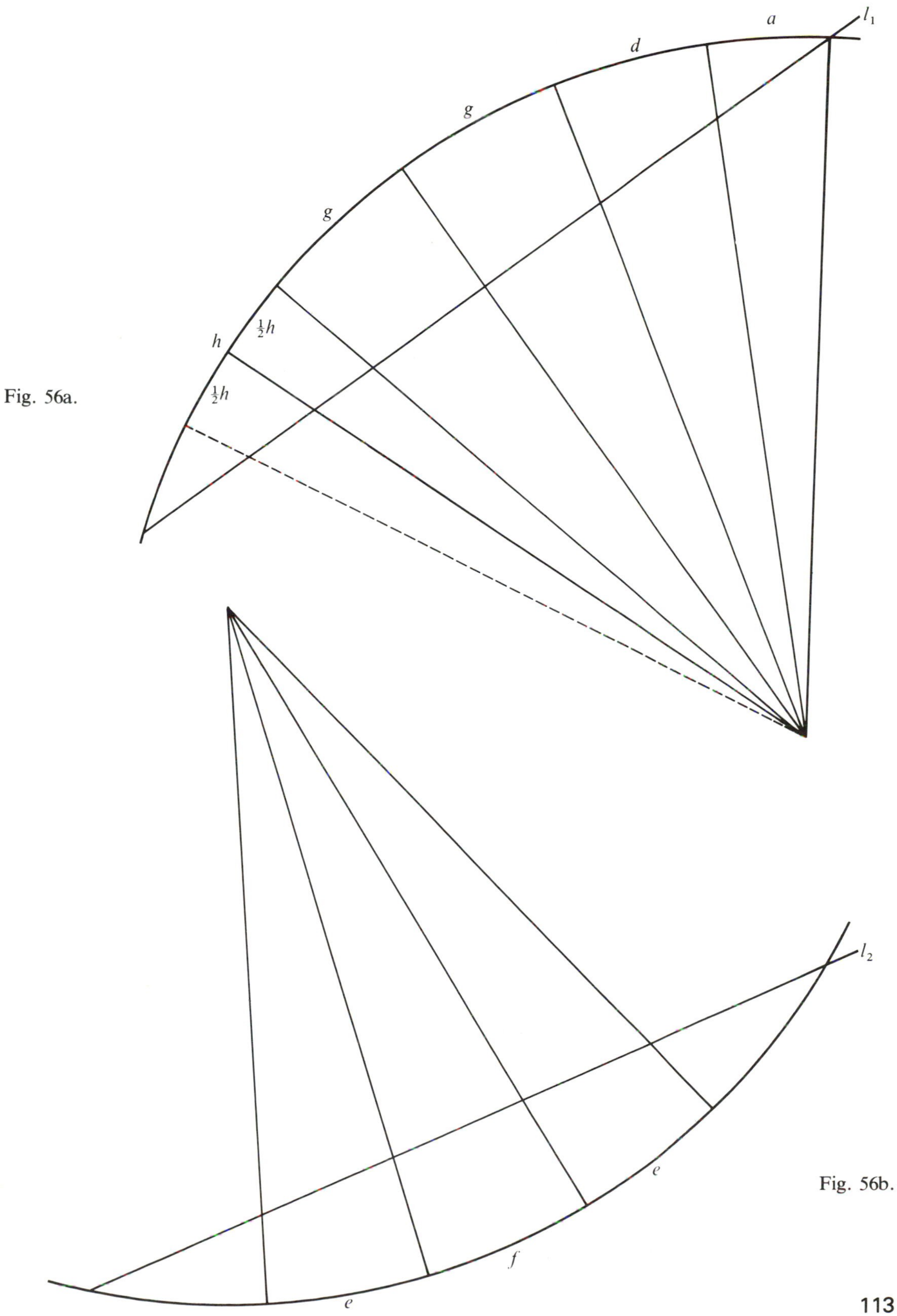

Fig. 56a.

Fig. 56b.

Fig. 56c.

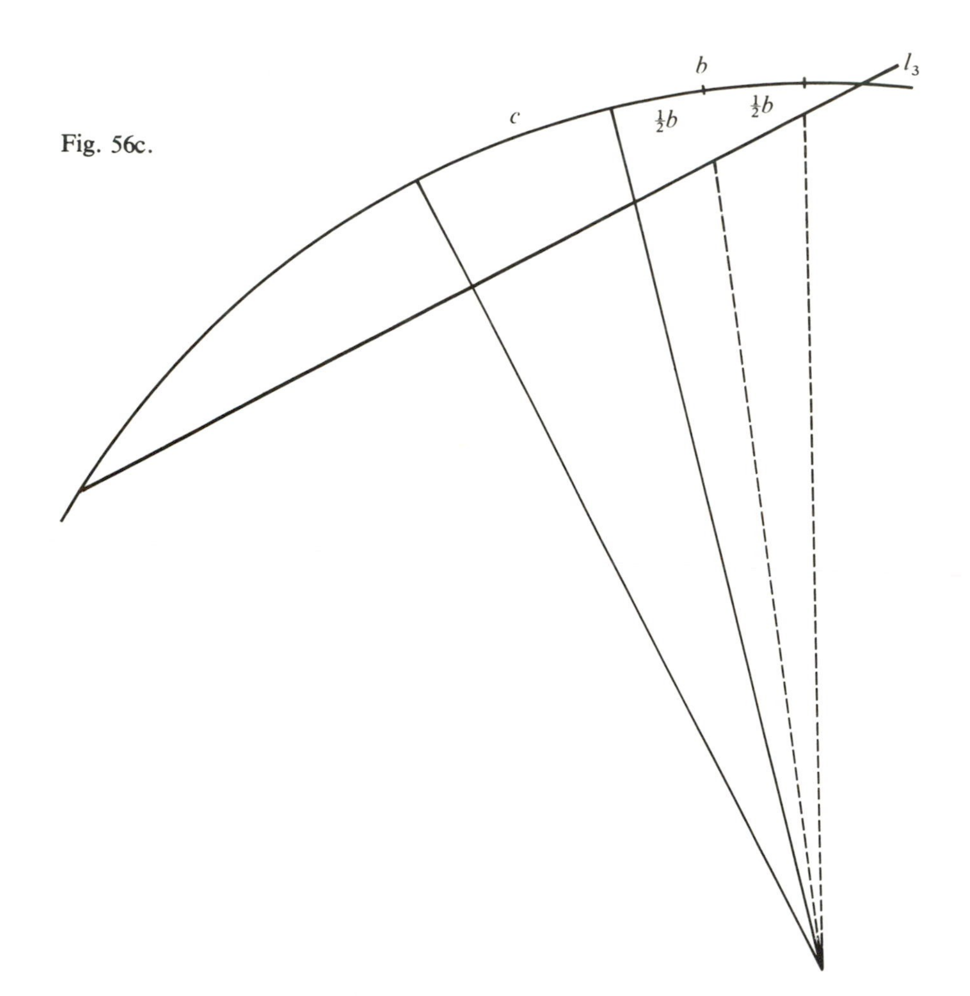

An alternative for geodesic segmentation of the snub dodecahedron

The spherical snub dodecahedron is $\{3, 5+\}_{2,1}$, already seen in Fig. 43. This too can have higher-frequency versions where $b = 2c$. Figure 57 shows the drawings for a model of $\{3, 5+\}_{6,3}$. For this model the method of equal arc segmentation along edges works very well. In Fig. 57, you see the chord l_1 is divided into segments derived from a trisection of the central angle at O_3. This must be done by measure of course, since every student of geometry well knows that the trisection of an angle in general cannot be done by purely geometrical or Euclidean construction alone. This makes the central linear segment of chord l_1 slightly longer than the two end segments. Transferring these lengths to the triangle face would introduce a "window" in the center of this face. But this is so slight it may be disregarded. In fact the projections relating to l_2 and l_3 lead to arcs so close to the same

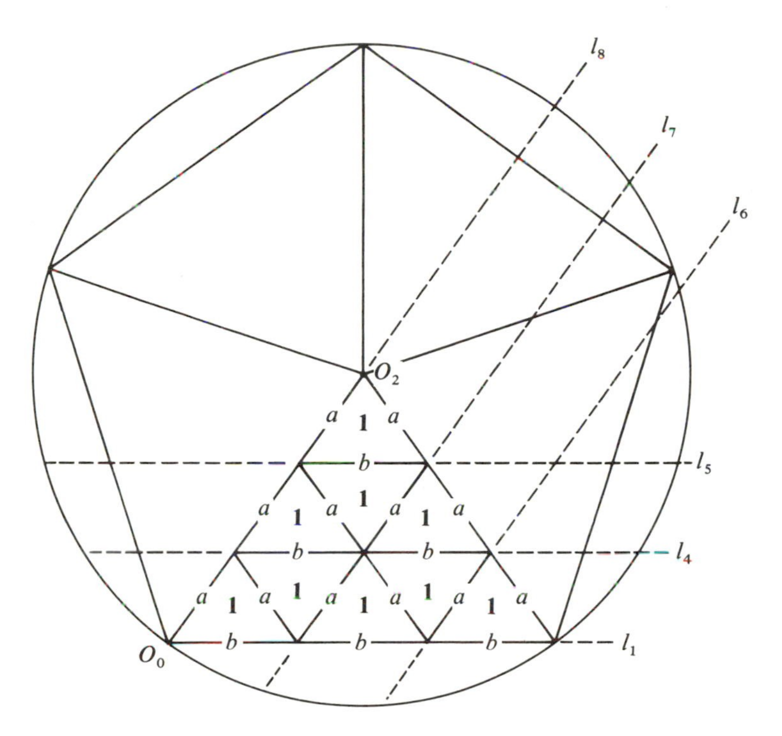

Fig. 57. A 9-frequency geodesic dome derived from the snub dodecahedron. $\{3, 5+\}_{6,3}$. (Figs. 57a–57g. Gnomonic projections.)

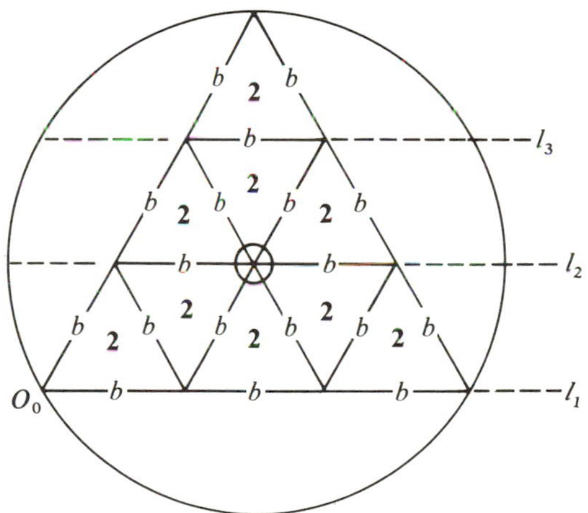

measure as those related to l_1, which has been named b, that no harm is done in making all arcs in the triangle face equal to b.

In Fig. 57b, chord l_8 shows point O_2 projected to Q_2 and then the arc O_0Q_2 is also trisected. The points of trisection on this arc are then joined by lines to the center of the sphere O_3. The points at which these lines cross chord l_8 determine the segments along O_0O_2 in the pentagon face. The 3-frequency grid in each face is then completed.

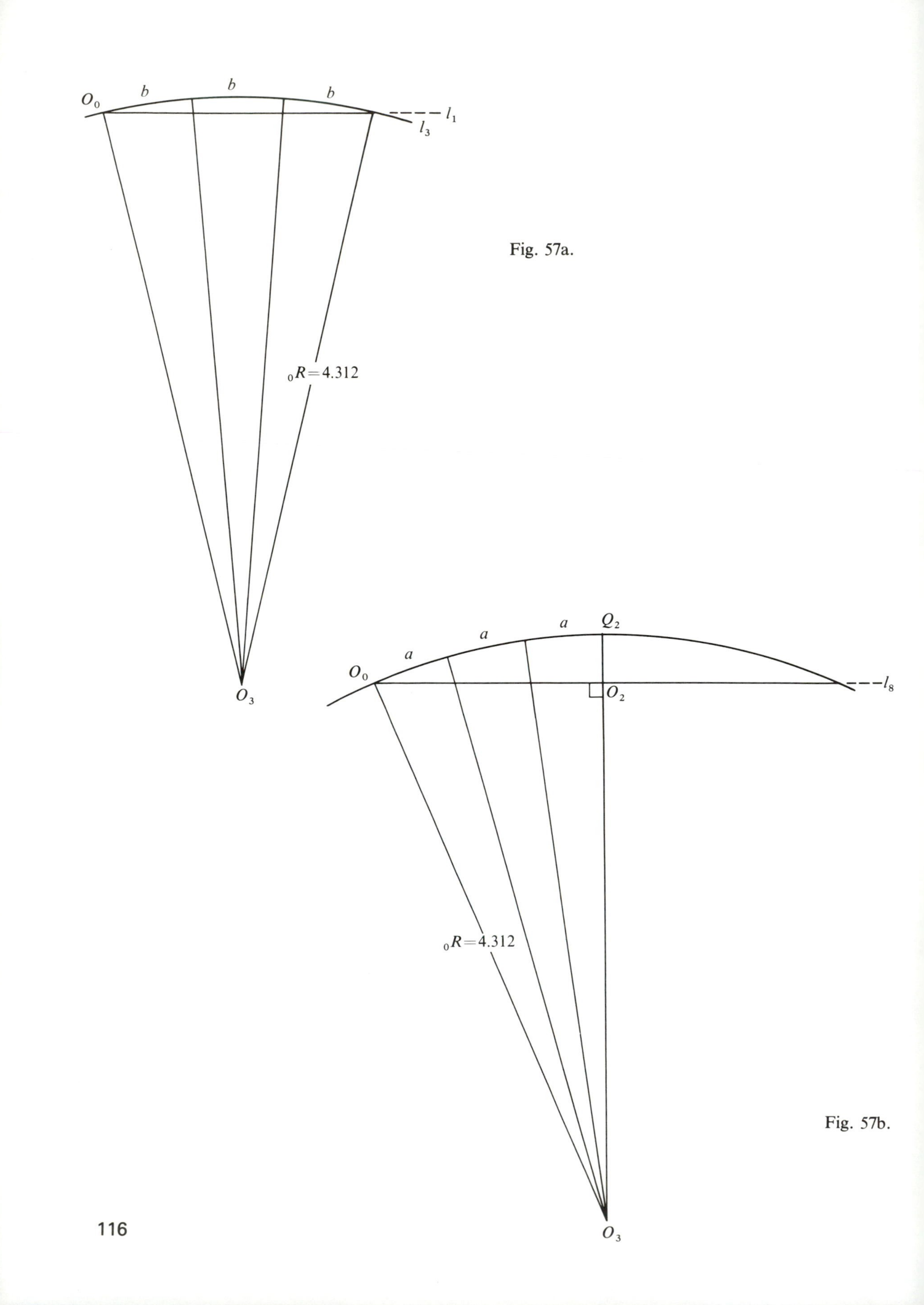

Fig. 57a.

Fig. 57b.

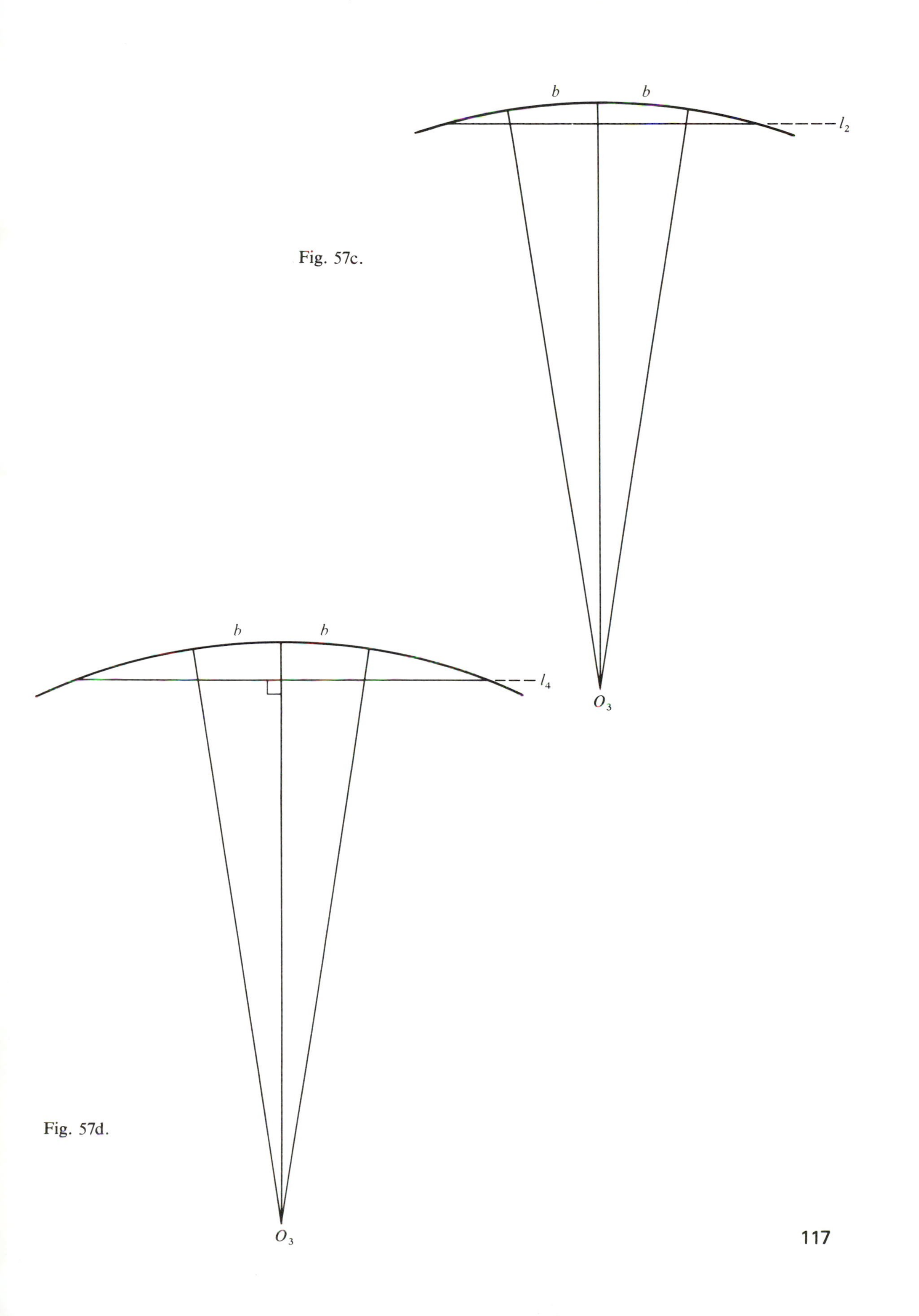

Fig. 57c.

Fig. 57d.

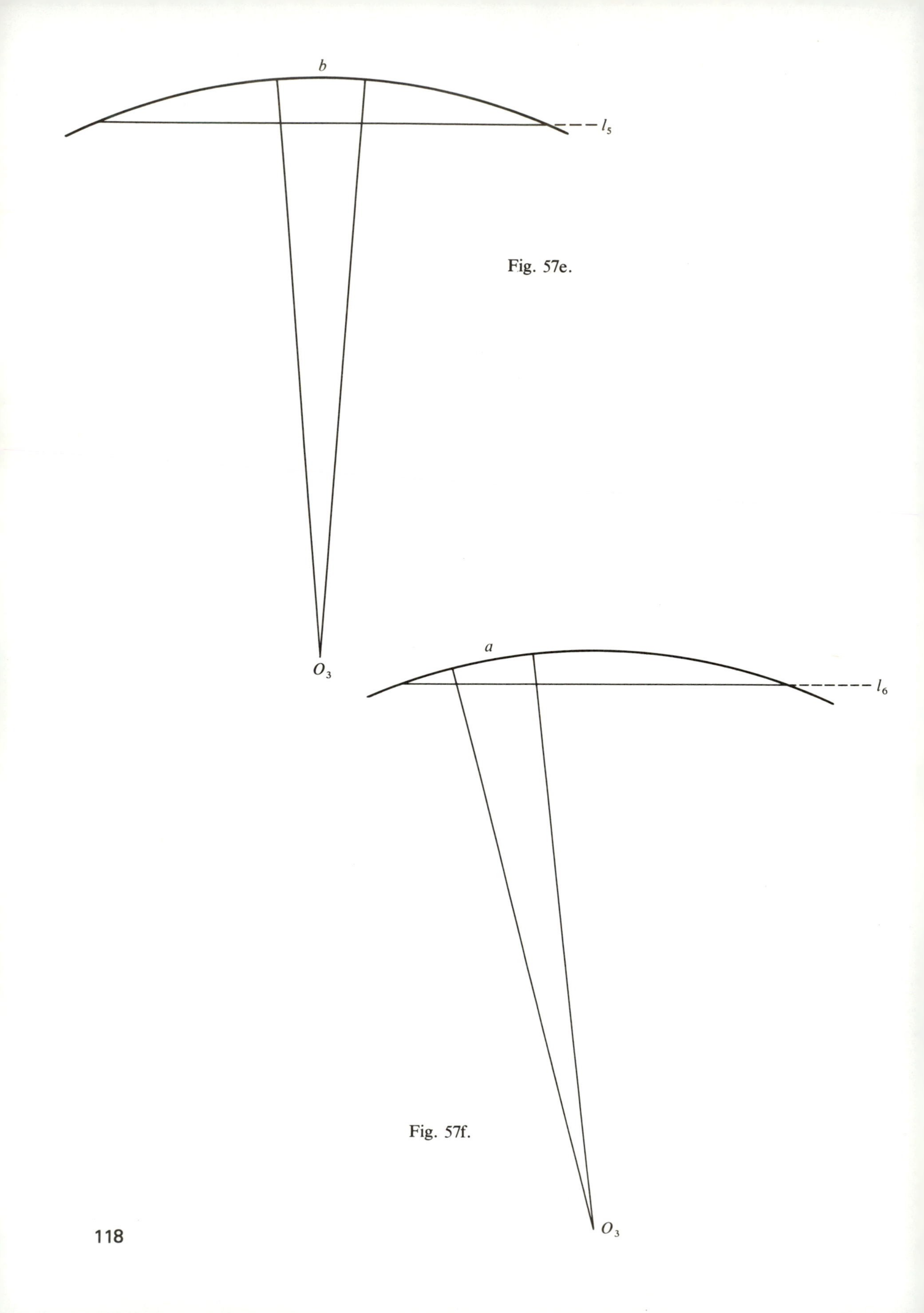

Fig. 57e.

Fig. 57f.

The gnomonic projections relating to l_4, l_5, l_6, l_7 are shown in Figs. 57d–57g. As you can see, these again lead to arcs so nearly equal that no harm is done in using only one further measure, namely a.

Now a very successful model can be made using only two types of bands. The assembly follows the right- or left-handed symmetry of the snub dodecahedron. The two arc measures and bands are as follows:

arcs	*bands*
a = 7.75 (approximately)	**1** $a\ b\ a$
b = 9.00 (approximately)	**2** $b\ b\ b$

Photo 43 shows the completed model. It contains 1,260 spherical triangles. This sets a record for the models in this book, having the least number of arc lengths and types of bands but giving a very highly complex geodesic dome – a 9-frequency dome.

Photo 43. $\{3, 5+\}_{6,3}$

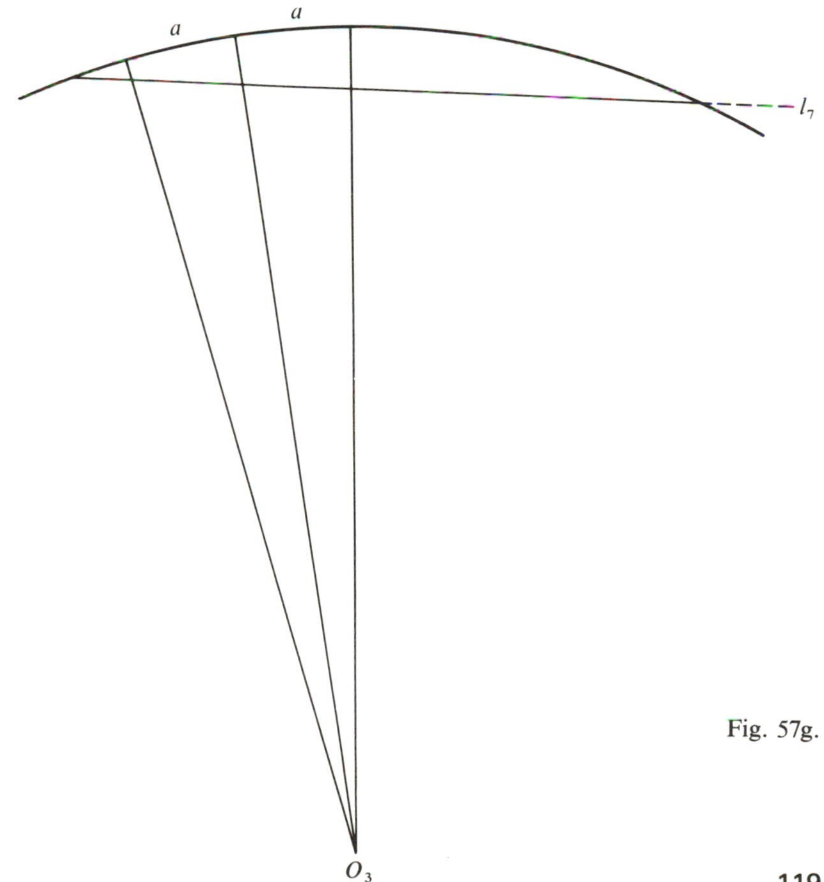

Fig. 57g.

A third alternative for geodesic segmentation of the icosahedron

It is possible to return to the icosahedral face for yet another type of tessellated network of smaller triangles leading to geodesic domes topologically equivalent to those derived from the snub dodecahedron. Figure 58 shows such a tessellation for a model of $\{3, 5+\}_{2,1}$; the projections are shown in Figs. 58a and 58b. Here too some of the arcs straddle the icosahedron edge, so the comment made with Fig. 54 applies here too. For this model, the arc measures and bands are as follows:

arcs	*bands*
$a = 22.692$	**1** *a b a*
$b = 26.212$	**2** *b c d*
$c = 27.802$	**3** *c c c*
$d = 27.016$	

For a model of $\{3, 5+\}_{4,2}$, see Fig. 59.

The gnomonic projections will be left for you to do as an exercise as well as the calculations, which are best done using the formulas of spherical trigonometry.

It is very fascinating to notice how the icosahedron is so adaptable in all of these cases. In fact the symbol $\{3, 5+\}_{b,c}$ can be seen as a simple way to arrange geodesic domes of the icosahedral symmetry group into three classes, as follows.

Class I Subscript b is an integer and c is always zero. The lines of segmentation on the flat icosahedral face run parallel to the edges.

Class II Subscript b is any integer and c is always equal to b. The lines of segmentation on the flat icosahedral face run perpendicular to the edges.

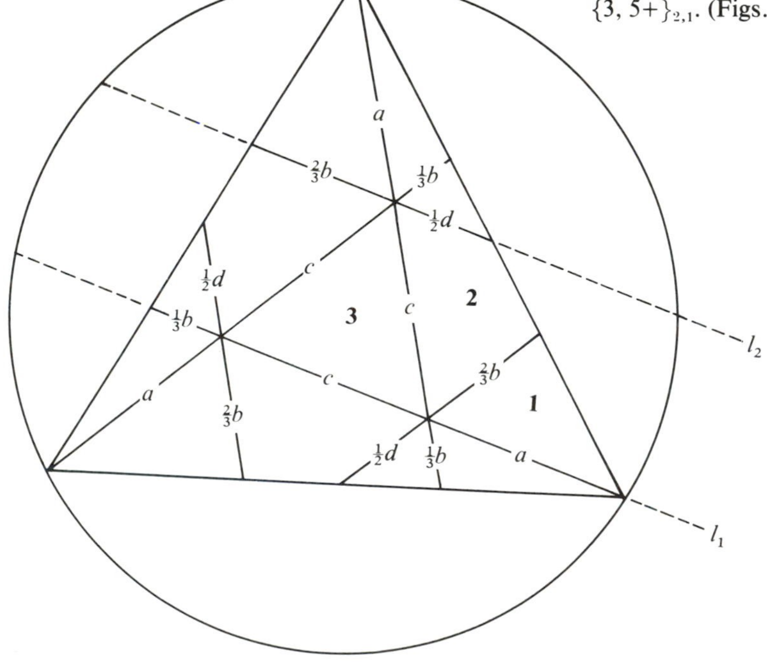

Fig. 58. A 3-frequency Class III icosahedron. $\{3, 5+\}_{2,1}$. (Figs. 58a, 58b. Gnomonic projections.)

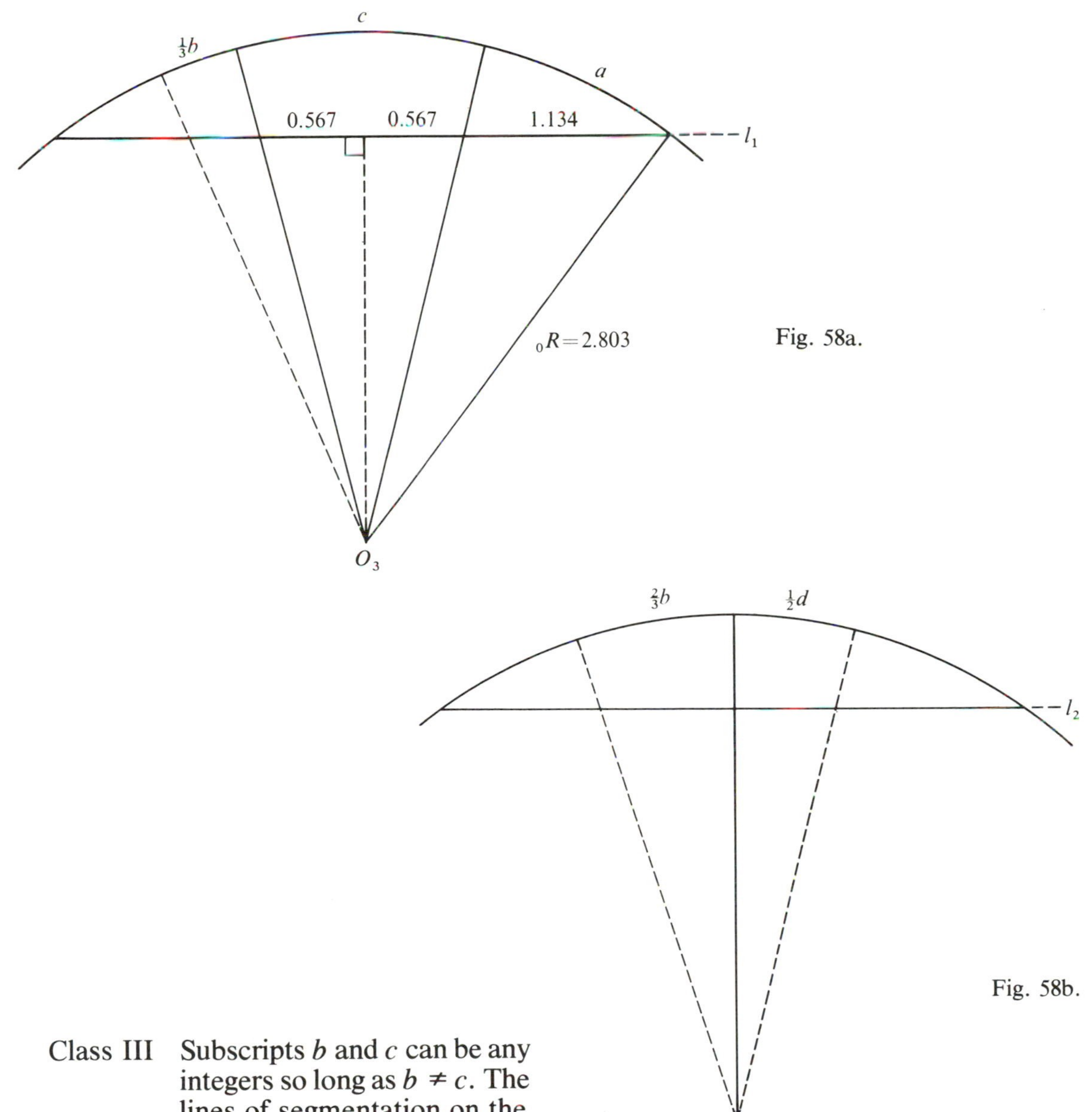

Fig. 58a.

Fig. 58b.

Class III Subscripts b and c can be any integers so long as $b \neq c$. The lines of segmentation on the flat icosahedral face run oblique to the edges.

Class I follows the symmetry of the icosahedron; class II follows the symmetry of the dodecahedron; and class III follows the symmetry of the snub dodecahedron. The frequency in all instances is $b + c$. You are invited to pursue these ideas further on your own, if you so desire.

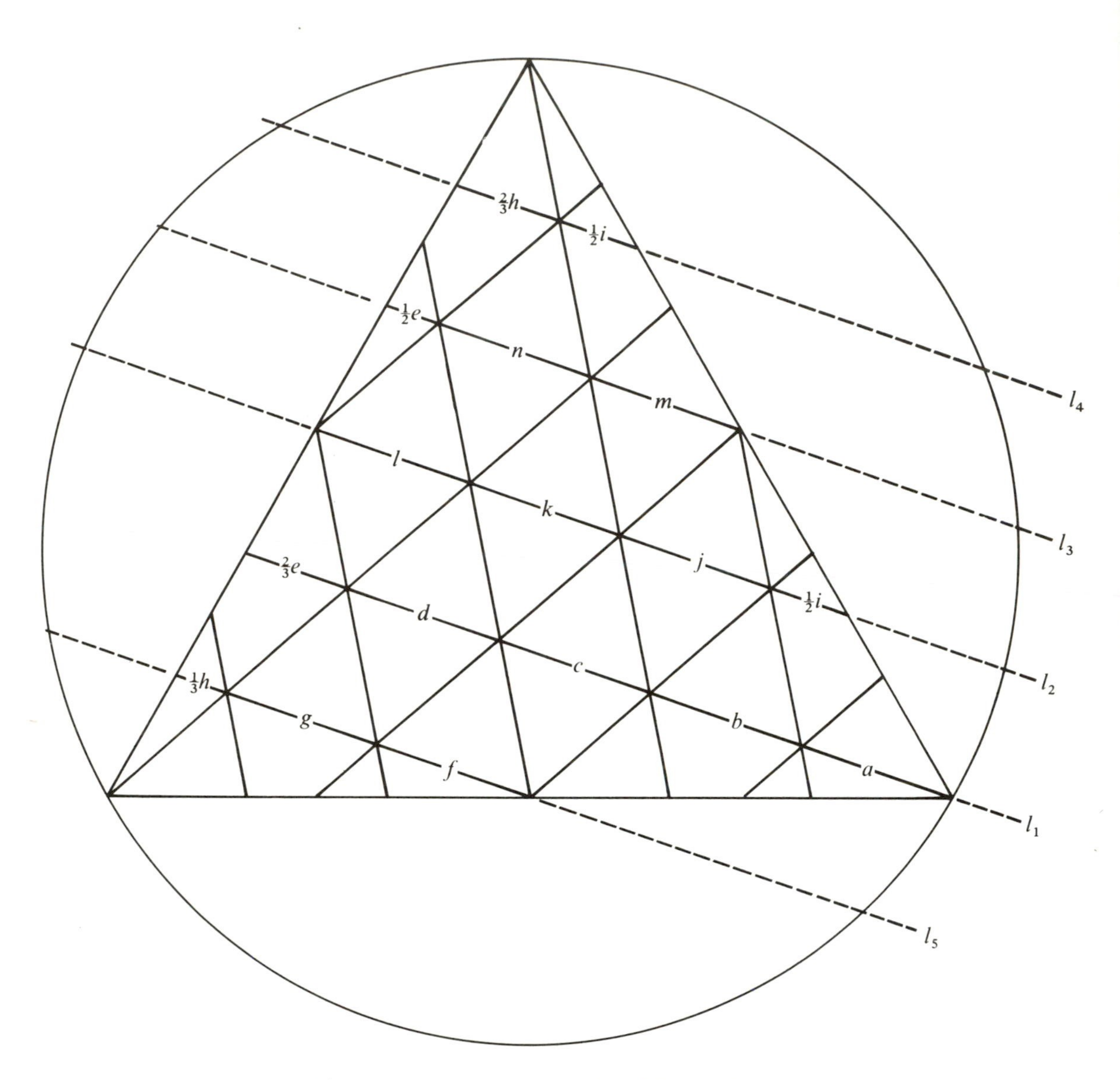

Fig. 59. A 6-frequency Class III icosahedron. $\{3, 5+\}_{4,2}$.

Final comments

Some final remarks are still worth making here. As you know, the snubs come in right- and left-handed versions. In the symbol this merely means an interchange of subscripts. For example, if $\{3, 5+\}_{6,3}$ is called right handed, then $\{3, 5+\}_{3,6}$ is left handed. The symbolism for the classification of geodesic domes can also be transformed into the dualized versions of these domes. For example, if $\{3, 5+\}_{b,c}$ is the original, its dual is $\{5+, 3\}_{b,c}$. The regular dodecahedron thus is $\{5+, 3\}_{1,0}$ and the truncated icosahedron is $\{5+, 3\}_{1,1}$. In these duals the faces are not tessellated

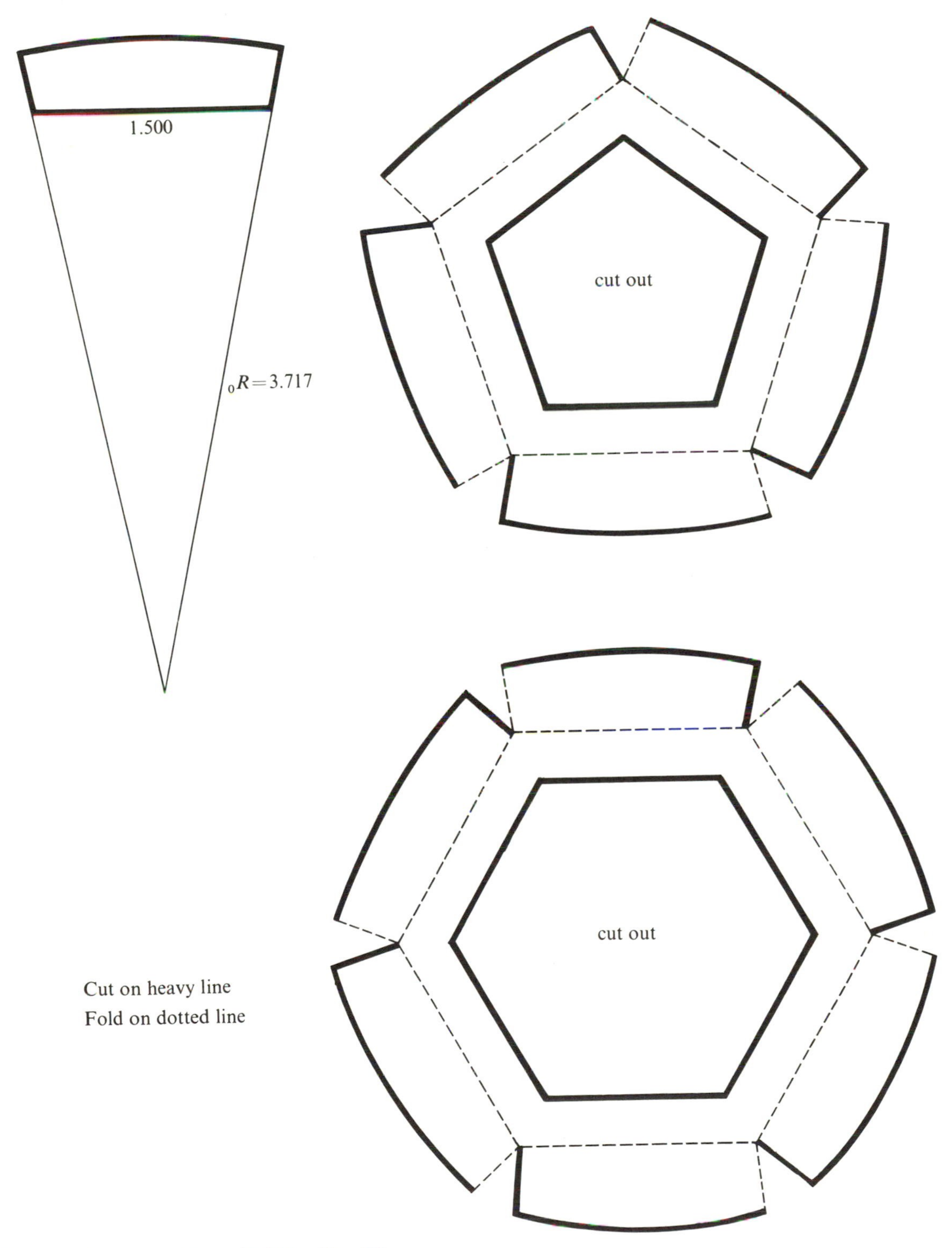

Fig. 60. A truncated icosahedron. $\{5+, 3\}_{1,1}$.

networks of triangles but rather pentagons and hexagons. They have 12 pentagon faces and $10(T - 1)$ hexagon faces. For this reason they do not readily lead to satisfactory models in paper. In the world of sports the soccer ball is a good example of $\{5+, 3\}_{1,1}$. A model of $\{5+, 3\}_{1,1}$ can be done in paper if the parts are designed as shown in Fig. 60.

The symbolism for geodesic domes in the icosahedral symmetry group can easily be transformed to fit the tetrahedral and octahedral symmetry groups as well. For the tetrahedral group, you would use $\{3, 3+\}_{b,c}$ and for the octahedral group, $\{3, 4+\}_{b,c}$. In these cases the formula $T = b^2 + bc + c^2$ remains valid. But for the tetrahedral group, $F = 4T$ and $E = 6T$; for the octahedral group, $F = 8T$ and $E = 12T$. The number of vertices V in both instances follows from the Euler formula. In the tetrahedral group, no matter what frequency the geodesic dome has, there will be 4 trivalent vertices whereas the rest are hexavalent. In the octahedral group, each geodesic dome will have 6 tetravalent vertices whereas the rest are hexavalent.

You can have a great deal of fun working out the higher frequencies for these and making some models following the same methods of geometrical construction and calculation as those used for the icosahedral group. You must see to it, however, that the appropriate radius ${}_0R$ is used in each case.

If you want some numerical data for these models, you can find the information in the books by Kenner and Pugh listed in the References at the end of this book.

V. Miscellaneous models

Consideration will now be given to various kinds of models that are closely related to the spherical models already presented in this book. After the very complex geodesic domes with which Section IV ended, you will no doubt be relieved to know that what follows will return to far simpler work.

Honeycomb models, edge models, and nolids

Look again at the regular spherical models with which this book began. The circular bands can be made with all the interior portions left inside. They can be made with chords taking the place of arcs, with or without the interior portions. Figures 61–63 illustrate the templates in these cases. When all the interior portions are left inside you get what might be called honeycomb models (see Fig. 64). All turn out to be rather delightful in appearance. You might say the old regular solids here take on a new look.

When some of the interior portion is removed the bands that are left for edge models, shown in Fig. 63, are not identically the same width, but for model making the differences are negligible. These edge models still retain the radii and apothems on the polyhedron faces. This seems to add to their beauty rather than to detract from it.

The regular polyhedrons can also be made with their circumscribing spheres. Figure 65 shows the layouts with the half-edge of the interior polyhedron shaded. This is best done for all parts in the model, so the interior polyhedron is clearly visible in relation to its circumscribing sphere.

With that much said for the regular solids, you can see that the same treatment can be given to the semiregular solids. All thirteen of them can be produced as honeycomb models. The arcs of the spherical bands are simply transformed into their respective chords, while χ, ϕ, ψ remain as central angles. If the interior portions are removed the radial lines automatically give you the correct angles for the folds that make all the parts fit properly for edge models.

Just as for the spherical models, so here too the work may be lessened by dropping a or r or both, but retaining e. This transforms all the orthoschemes into pyramidal cells: triangular, square, pentagonal, hexagonal, octagonal, and decagonal. These pyramidal cells are cemented to one another so that lateral faces coincide and all the vertices of the pyramids point inward to the original center of the circumscribing sphere. The fact that the lateral faces are all triangles reintroduces sufficient rigidity so that the cube, the dodecahedron, or any of the semiregular polyhedrons come out as satisfactory models. When these are made using glossy or metalic paper they become very attractive ornamental devices.

Honeycomb models in some cases become what Holden, in *Shapes, space and symmetry*, pp. 124–5, calls nolids. Regular nolids are composed of regular polygons all passing through the center of the nolid. The octahedron can be made to show its diametral squares, the cuboc-

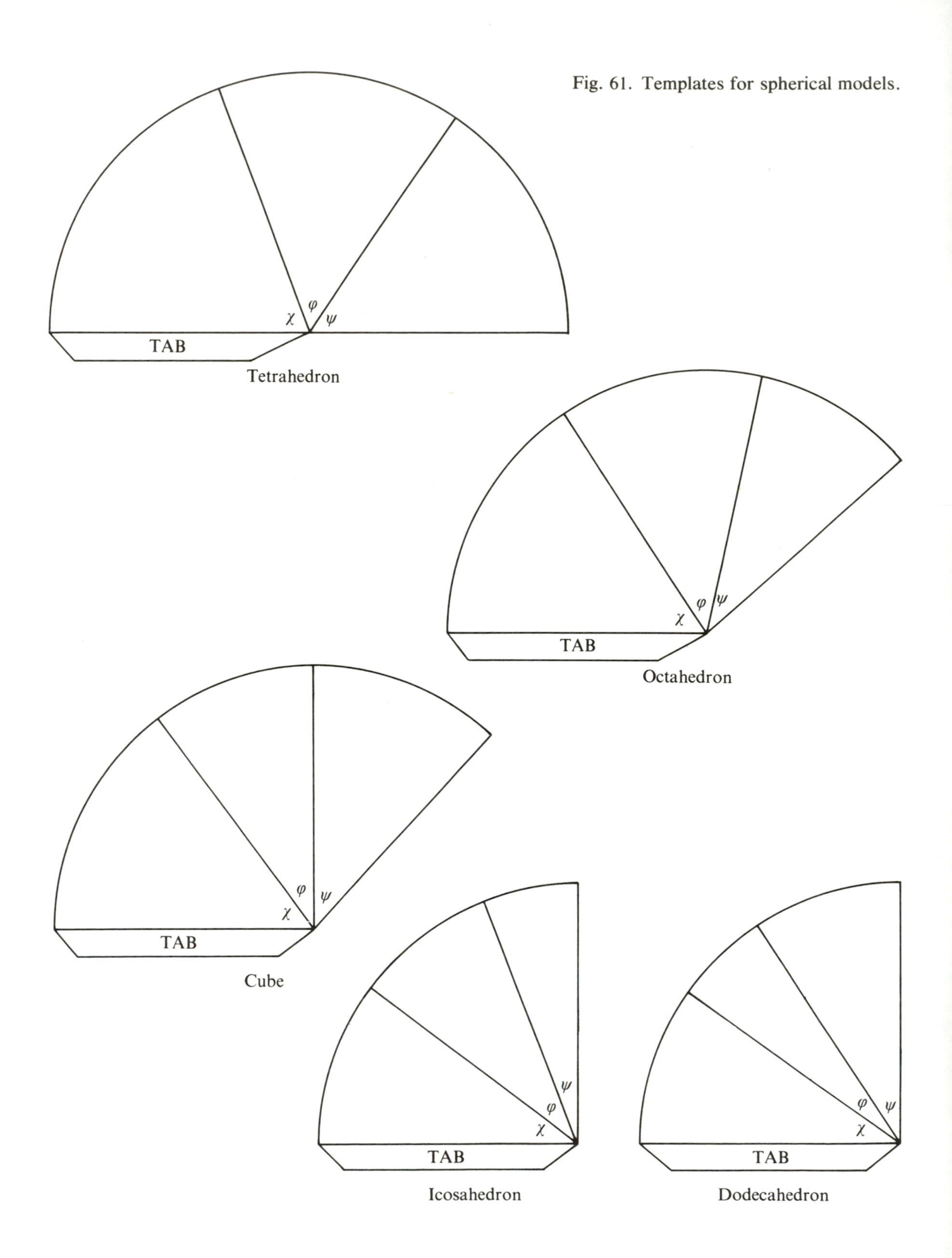

Fig. 61. Templates for spherical models.

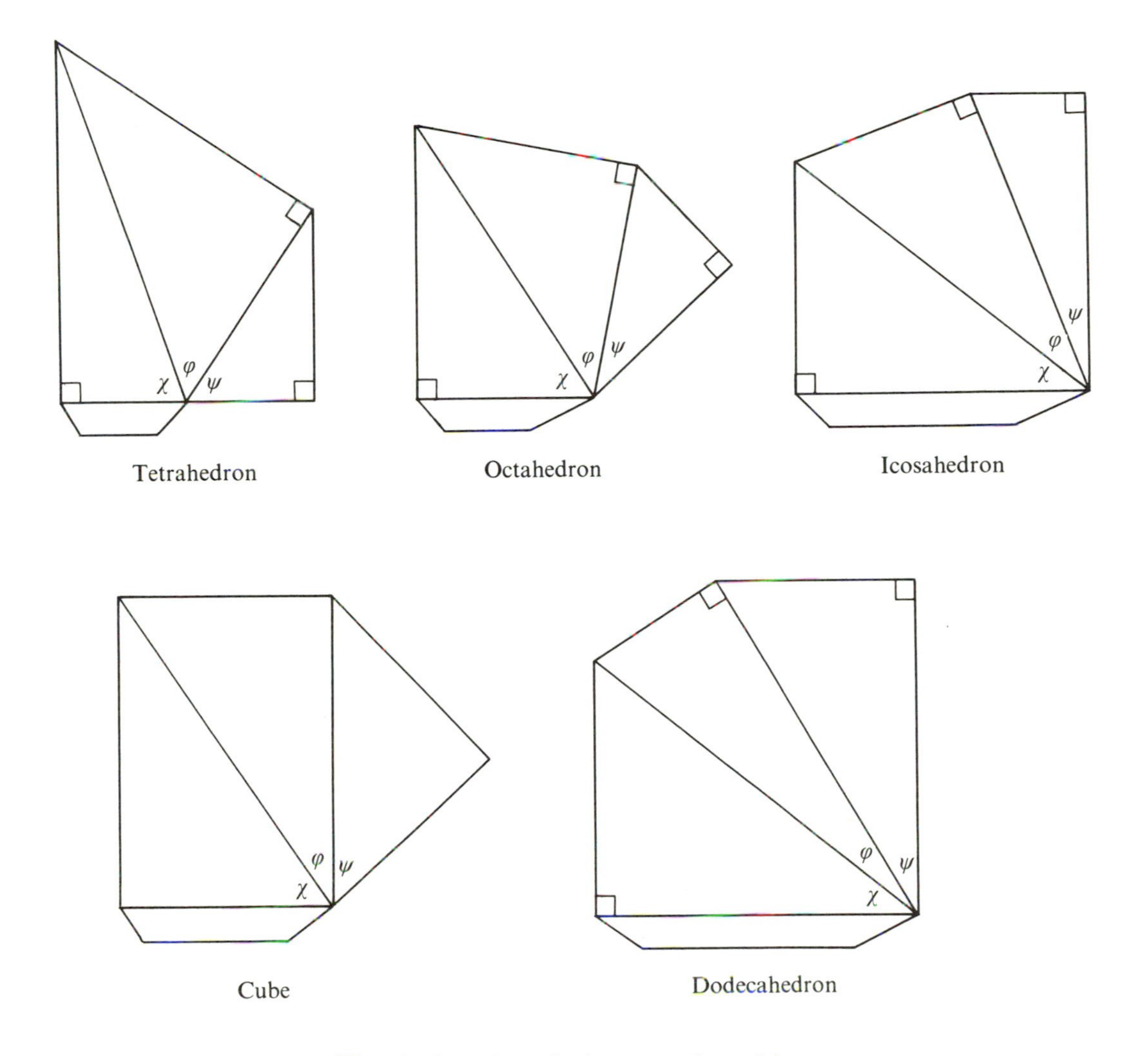

Fig. 62. Templates for honeycomb models.

tahedron its diametral hexagons, and the icosidodecahedron its diametral decagons (see Fig. 66). You should be able to design your own templates for these including the tabs for pyramidal parts. These tabs then disappear between the cemented parts. The double thickness of paper gives added rigidity.

A particularly pleasing nolid can be made from the layouts in Fig. 67. It has ten diametral hexagons that intersect in such a way that their edges form twelve pentagrams. The method of assembly is the same as that suggested for Fig. 66. If the interior parts of the layouts in Fig. 67 are removed and the edges changed into circular bands, the result is a model with ten great circles intersecting to form spherical pentagrams on the surface of a sphere.

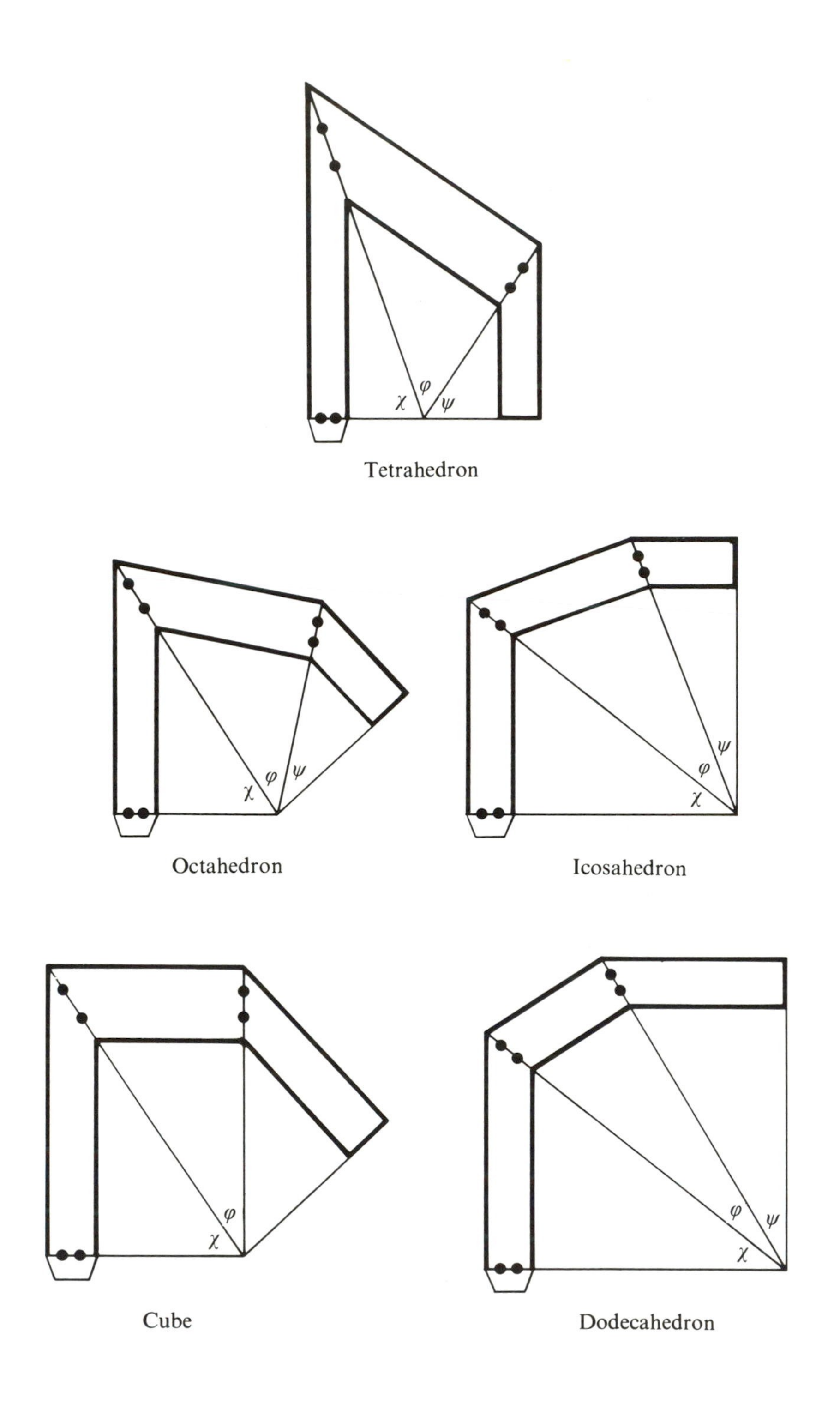

Fig. 63. Templates for edge models.

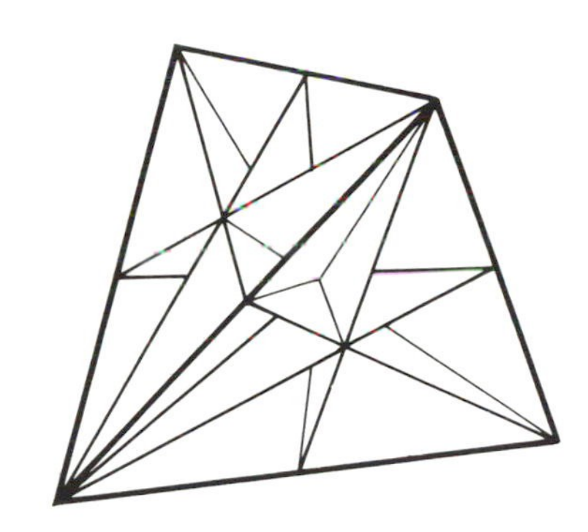

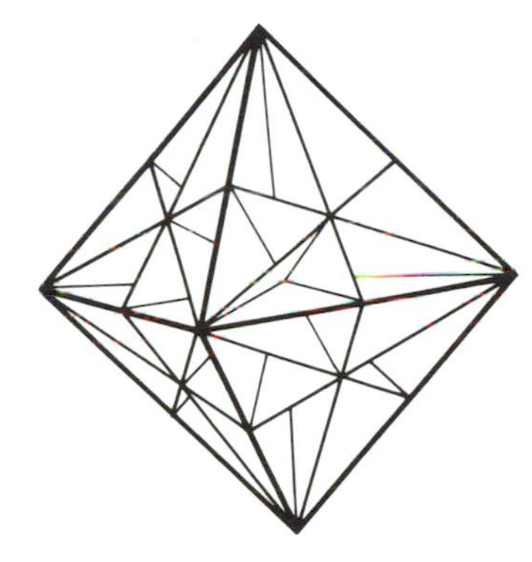

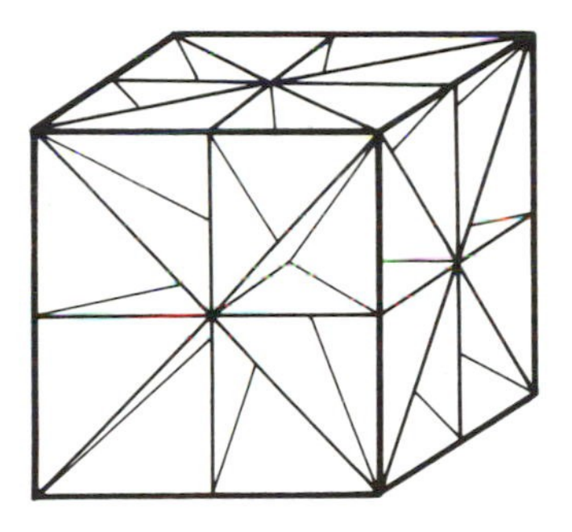

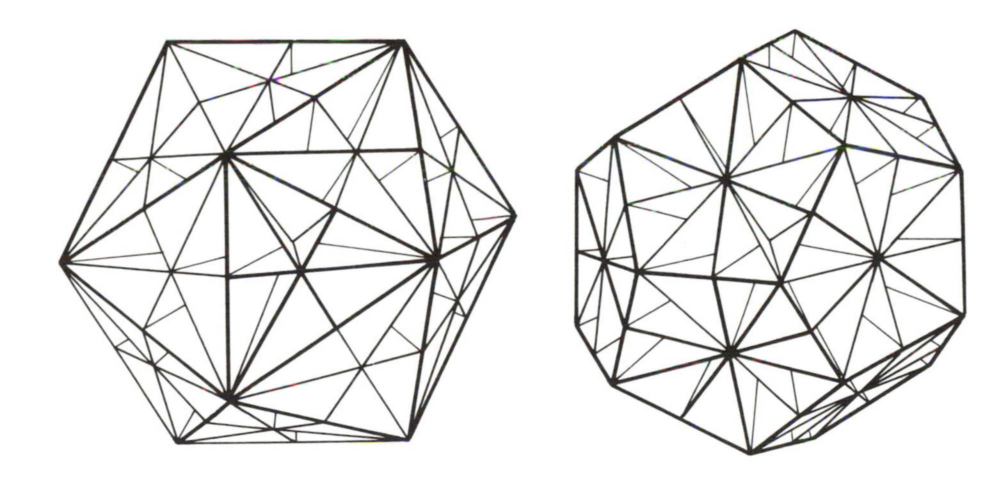

Fig. 64. The five regular solids as honeycomb models.

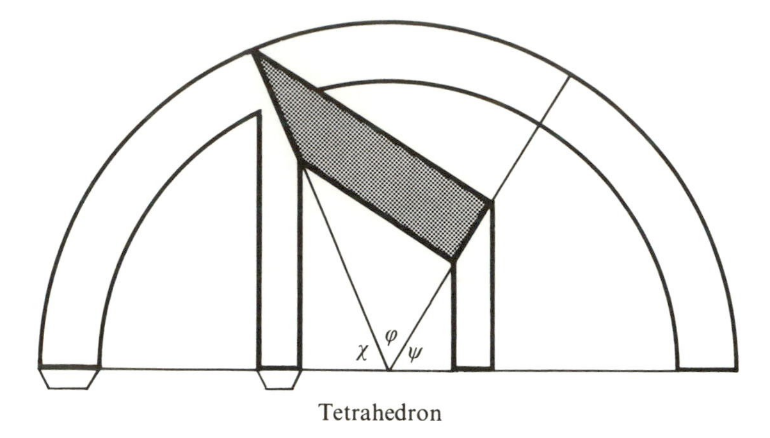

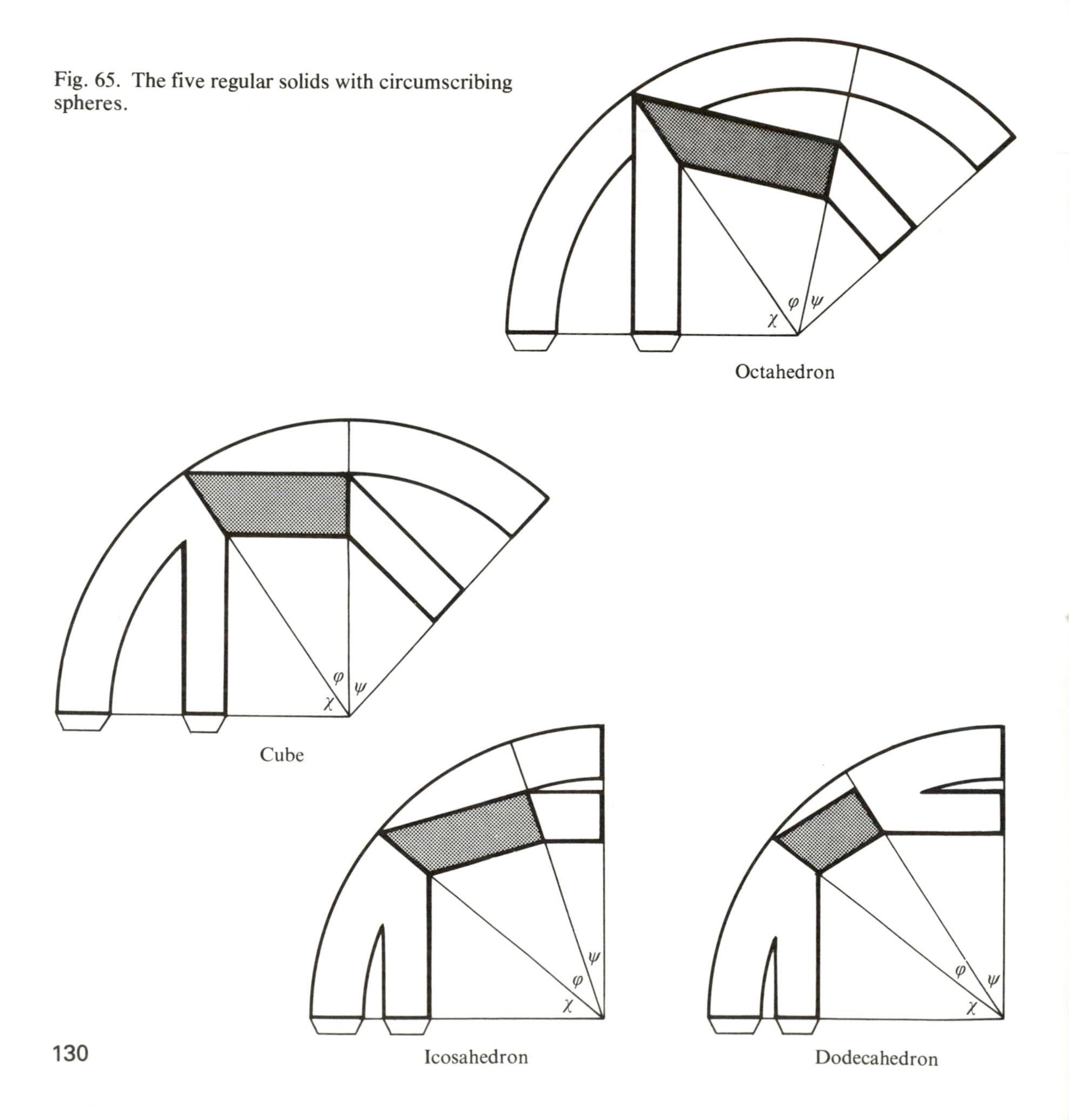

Fig. 65. The five regular solids with circumscribing spheres.

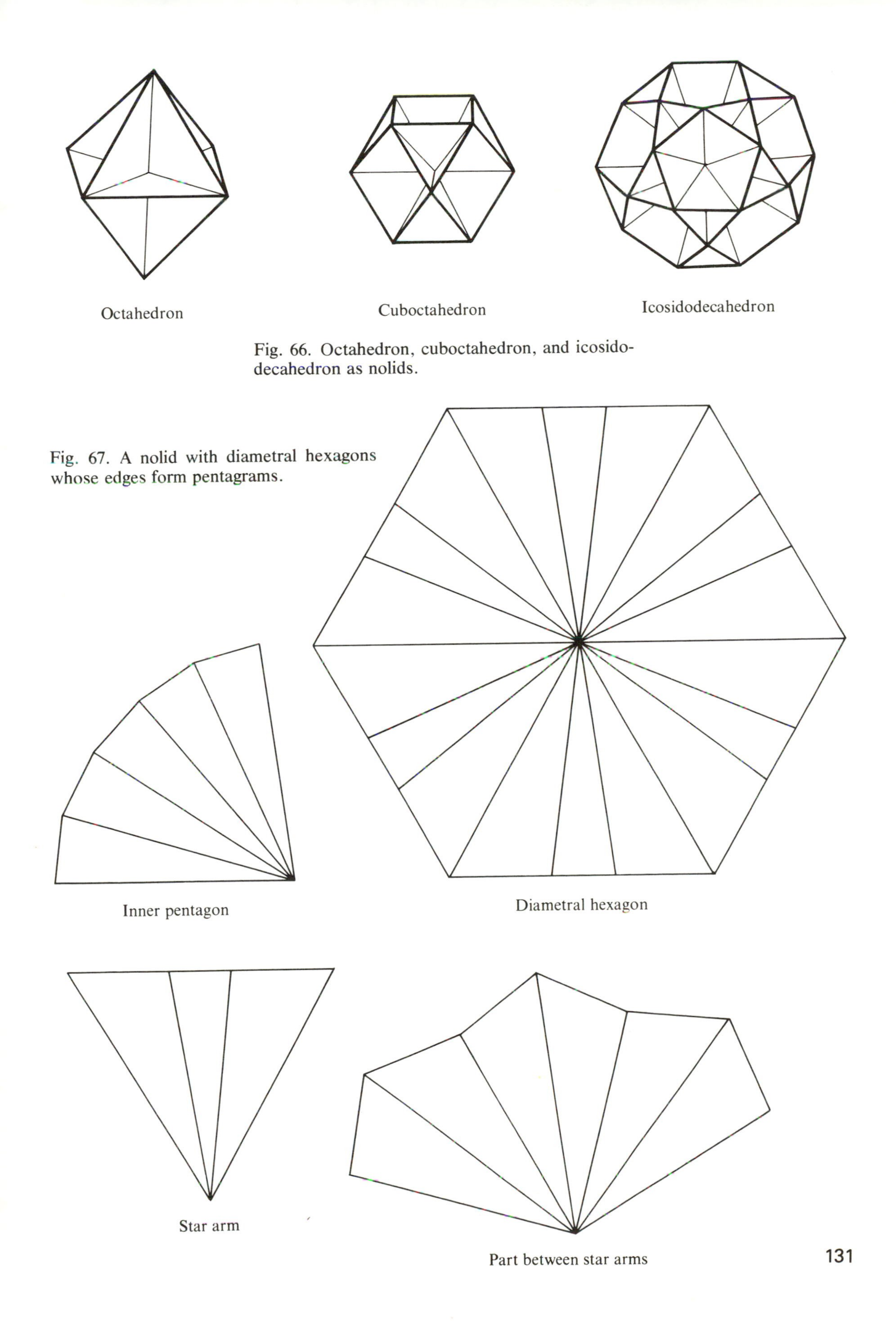

Fig. 66. Octahedron, cuboctahedron, and icosidodecahedron as nolids.

Fig. 67. A nolid with diametral hexagons whose edges form pentagrams.

An introduction to the notion of polyhedral density

In Section III spherical models derived from star-faced polyhedrons were mentioned, but only convex polyhedrons with star patterns on their facial planes served as basic polyhedrons for spherical models. Some of these models seem to have a close relationship to nonconvex uniform polyhedrons, many of which do indeed have pentagrams, octagrams, and decagrams as faces. You may have been wondering why the regular stellated polyhedrons were never mentioned: the small stellated dodecahedron, the great dodecahedron, the great stellated dodecahedron, and the great icosahedron. The reason is that projecting any one of these onto its circumscribing sphere will cover the sphere more than once. There is no simple way of showing this in a model. It is a mathematical abstraction, like coincident points, lines, and planes in geometry. Here the planes involve a partial and multiple covering of the surface of the sphere. This leads to what is called polyhedral density. The three regular spherical models of Plates 9–11 may be thought of as spherical tessellations covering the surface of the sphere once. The characteristic triangles are also called Möbius triangles, because it was Möbius (1849) who first observed this. So-called Schwarz triangles cover the sphere more than once but still only a finite number of times. It was Schwarz (1873) who first listed the total number of ways in which this can be done.

It may be very difficult to image multiple coverings of the sphere, but Coxeter has given a very simple way of locating at least one Schwarz triangle, which uses the fact that a Schwarz triangle can be decomposed into a set of Möbius triangles. In a Schwarz triangle $(p\ q\ r)$, one or more of p, q, r may be rational; that is, certain fractions may be used as replacements for p, q, r. Coxeter's symbol for the small stellated dodecahedron is $5 \mid 2\ {}^5/_2$. What this means is that you can take the icosahedral model with its 120 characteristic triangles covering the surface of the sphere once and pick out from among these triangles a subgroup composed of only three characteristic triangles arranged so that they form another spherical triangle $(5\ 2\ {}^5/_2)$. Such a Schwarz triangle is shaded in Fig. 68a. Figures 68b and 68c show two other examples, one from the octahedral group and the other from the tetrahedral. The $\mid$ in Coxeter's symbol for a uniform polyhedron indicates the position of the point P whose images along with P mark the vertices of the polyhedron in question.

For the small stellated dodecahedron, point P is in the position shown in Fig. 68d. Here any one of the three characteristic triangles can occupy any one of the three positions inside the larger triangle. This means that the sphere is covered three times by any one of them. The small stellated dodecahedron thus has a density of 3; the great dodecahedron also has a density of 3.

The great stellated dodecahedron and the great icosahedron have a density of 7. This is illustrated in the grouping of characteristic triangles shown in Fig. 68e. Notice that the number of characteristic triangles is equal to the density.

You can now use the three regular spherical models that you have made to find an example of each of the Schwarz triangles listed in Table 7.

As has already been said, the $\mid$ in Coxeter's symbols for uniform polyhedrons indicates the position of point P whose images along with P generate the vertices and hence also the edges and faces in all the convex and nonconvex polyhedrons. Locating the faces in the convex cases is not too difficult. In the nonconvex cases it is a real challenge. You will do well if you

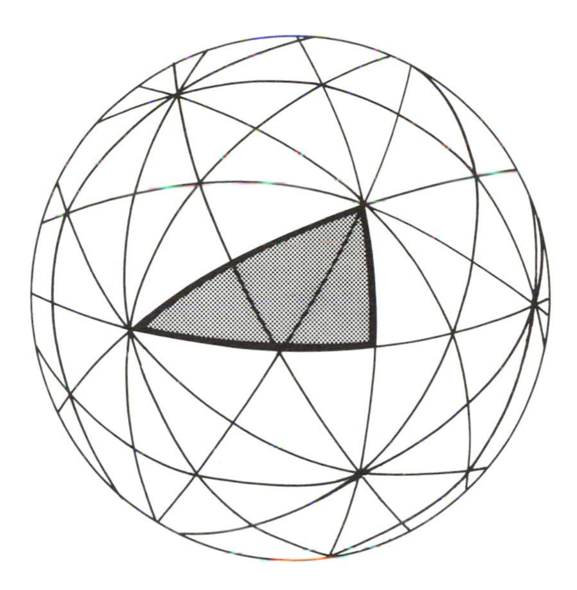

$(2\ \frac{5}{2}\ 5)$

density 3

Fig. 68a.

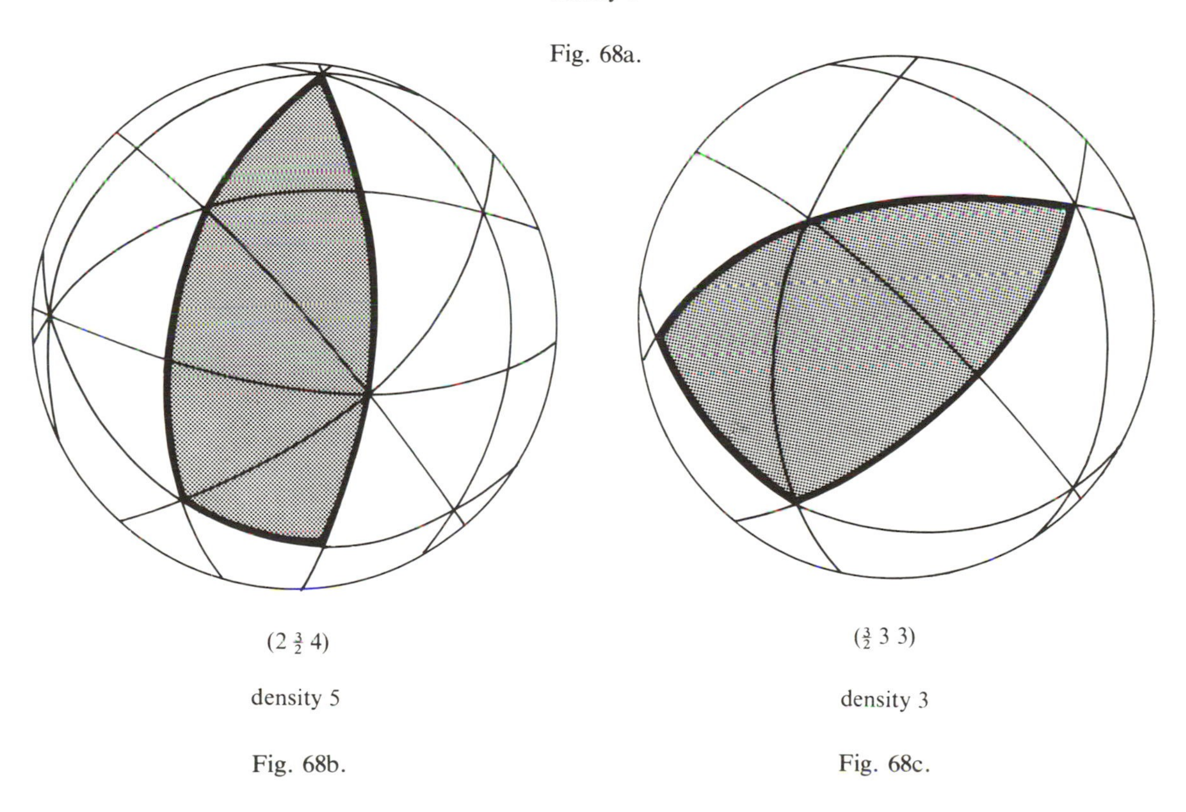

$(2\ \frac{3}{2}\ 4)$

density 5

Fig. 68b.

$(\frac{3}{2}\ 3\ 3)$

density 3

Fig. 68c.

Figs. 68a–68c. Examples of polyhedral density.

can pick out one or the other star faces in any of the nonconvex uniform polyhedrons. To see all the intersecting facial planes does not seem to come within the range of the imaginative powers of the average person.

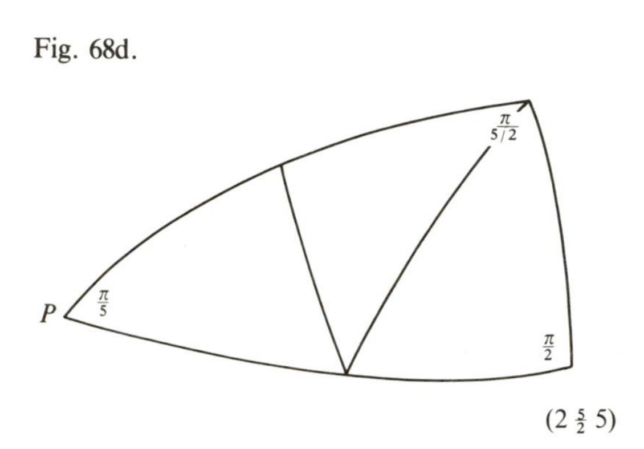

Fig. 68d.

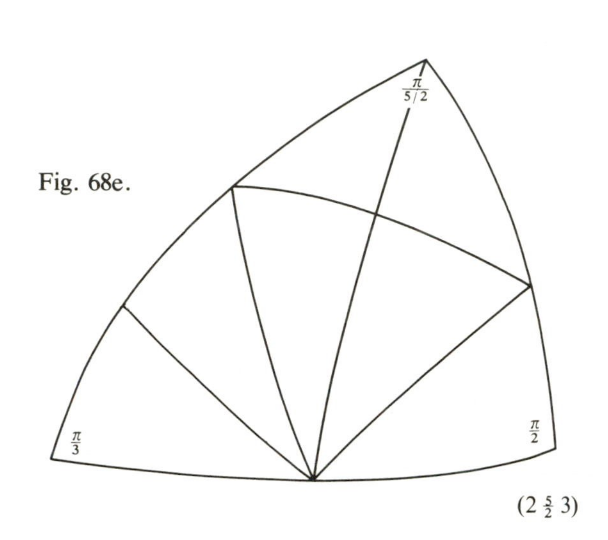

Fig. 68e.

Table 7. *Schwarz triangles*

In the tetrahedral model:	In the octahedral model:
$(2\ 3\ 3)$	$(2\ 3\ 4)$
$(2\ ^3/_2\ 3)$	$(2\ ^3/_2\ 4)$
$(2\ ^3/_2\ ^3/_2)$	$(2\ 3\ ^4/_3)$
$(^3/_2\ 3\ 3)$	$(2\ ^3/_2\ ^4/_3)$
$(^3/_2\ ^3/_2\ ^3/_2)$	$(^3/_2\ 4\ 4)$
	$(3\ ^4/_3\ 4)$
	$(^3/_2\ ^4/_3\ ^4/_3)$

In the icosahedral model:

$(2\ 3\ 5)$	$(2\ ^5/_2\ 5)$	$(^3/_2\ 3\ 5)$
$(2\ ^3/_2\ 5)$	$(2\ ^5/_3\ 5)$	$(3\ 3\ ^5/_4)$
$(2\ 3\ ^5/_4)$	$(2\ ^5/_2\ ^5/_4)$	$(^3/_2\ ^3/_2\ ^5/_4)$
$(2\ ^3/_2\ ^5/_4)$	$(2\ ^5/_3\ ^5/_4)$	$(^5/_4\ 5\ 5)$
$(^5/_2\ 3\ 3)$	$(^5/_3\ 3\ 5)$	$(^5/_4\ ^5/_4\ ^5/_4)$
$(^5/_3\ ^3/_2\ 3)$	$(^5/_2\ ^3/_2\ 5)$	$(2\ ^5/_2\ 3)$
$(^5/_2\ ^3/_2\ ^3/_2)$	$(^5/_2\ 3\ ^5/_4)$	$(2\ ^5/_3\ 3)$
$(^3/_2\ 5\ 5)$	$(^5/_3\ ^3/_2\ ^5/_4)$	$(2\ ^5/_2\ ^3/_2)$
$(3\ ^5/_4\ 5)$	$(^5/_2\ ^5/_2\ ^5/_2)$	$(2\ ^5/_3\ ^3/_2)$
$(^3/_2\ ^5/_4\ ^5/_4)$	$(^5/_3\ ^5/_3\ ^5/_2)$	$(^5/_3\ ^5/_2\ 3)$
		$(^5/_2\ ^5/_2\ ^3/_2)$
		$(^5/_3\ ^5/_3\ ^3/_2)$

Fig. 69. Edge models of the small stellated dodecahedron and the great stellated dodecahedron. ▷

Edge models of stellated forms

If spherical models cannot be made of the regular stellated polyhedrons, variations can be made that will yield chord lengths leading to edge models. The paper bands needed for the small stellated dodecahedron and the great stellated dodecahedron are shown in Fig. 69. Leaving the interior portions would lead to honeycomb models of these stellated forms, all of which are very attractive.

The great dodecahedron and the great icosahedron are not so readily made in this manner because their faces intersect along recessed lines. The small stellated dodecahedron, however, has the same edges as the great icosahedron, and the great dodecahedron has the same edges as a regular icosahedron. Such facts are helpful in designing templates for models. It will be left for your investigation should you have any further interest. Photos 44–47 show the models.

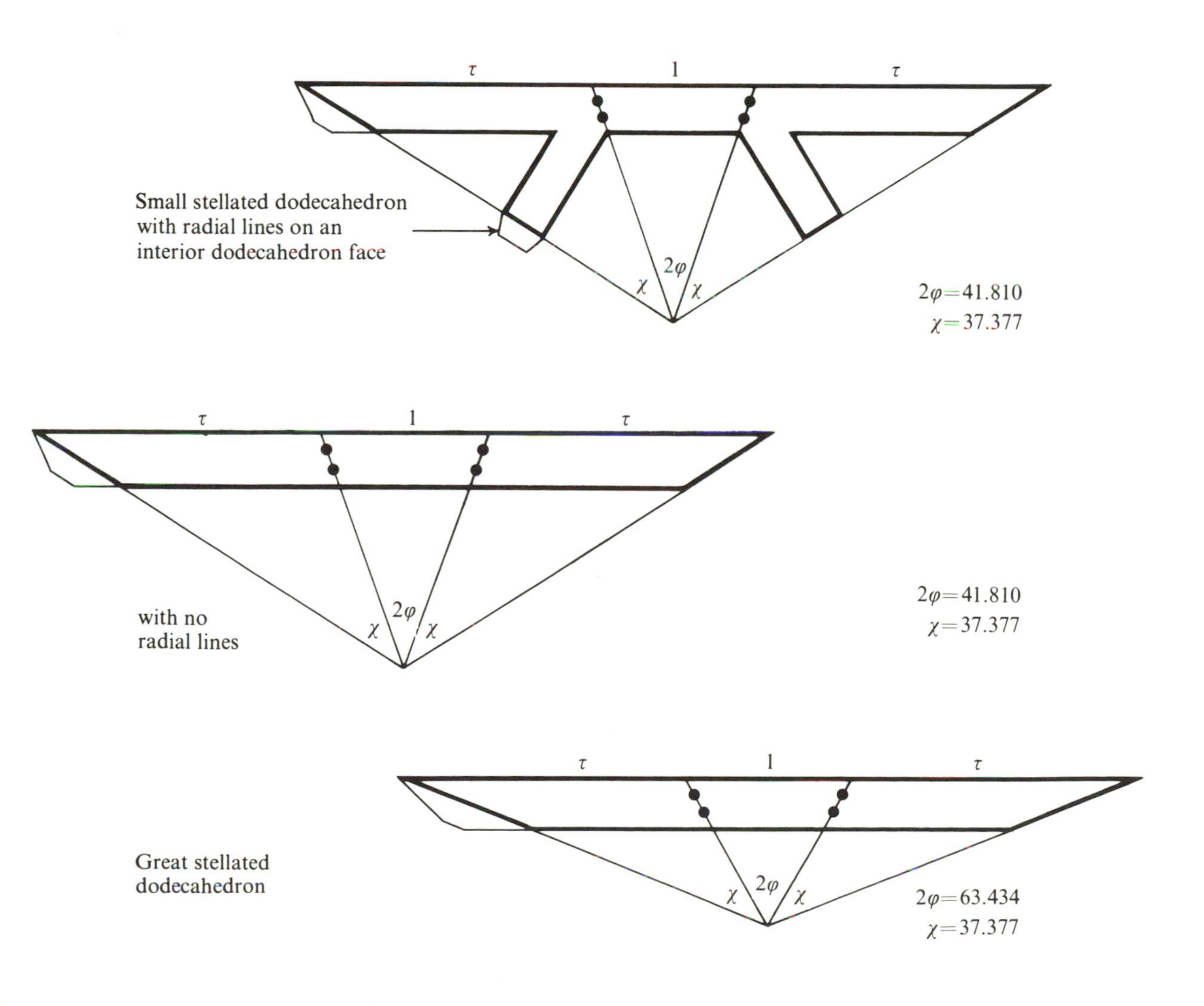

Photo 44. Dodecahedron.

Photo 45. Small stellated dodecahedron (with interior dodecahedron).

Photo 46. Great icosahedron.

Photo 47. Great stellated dodecahedron.

Some final comments about geodesic domes

The spherical models presented in this book were all truly shperical in the sense that the arcs were all great circle arcs. You may perhaps be aware that in large-scale structures designed by engineers and architects, geodesic domes look spherical but actually they are built using straight struts, so that the triangles are all flat triangles made of opaque or translucent material or metal sheets, whereas the struts can be of wood or steel. If you want to make small-scale models like these in paper, you can see that slight variations

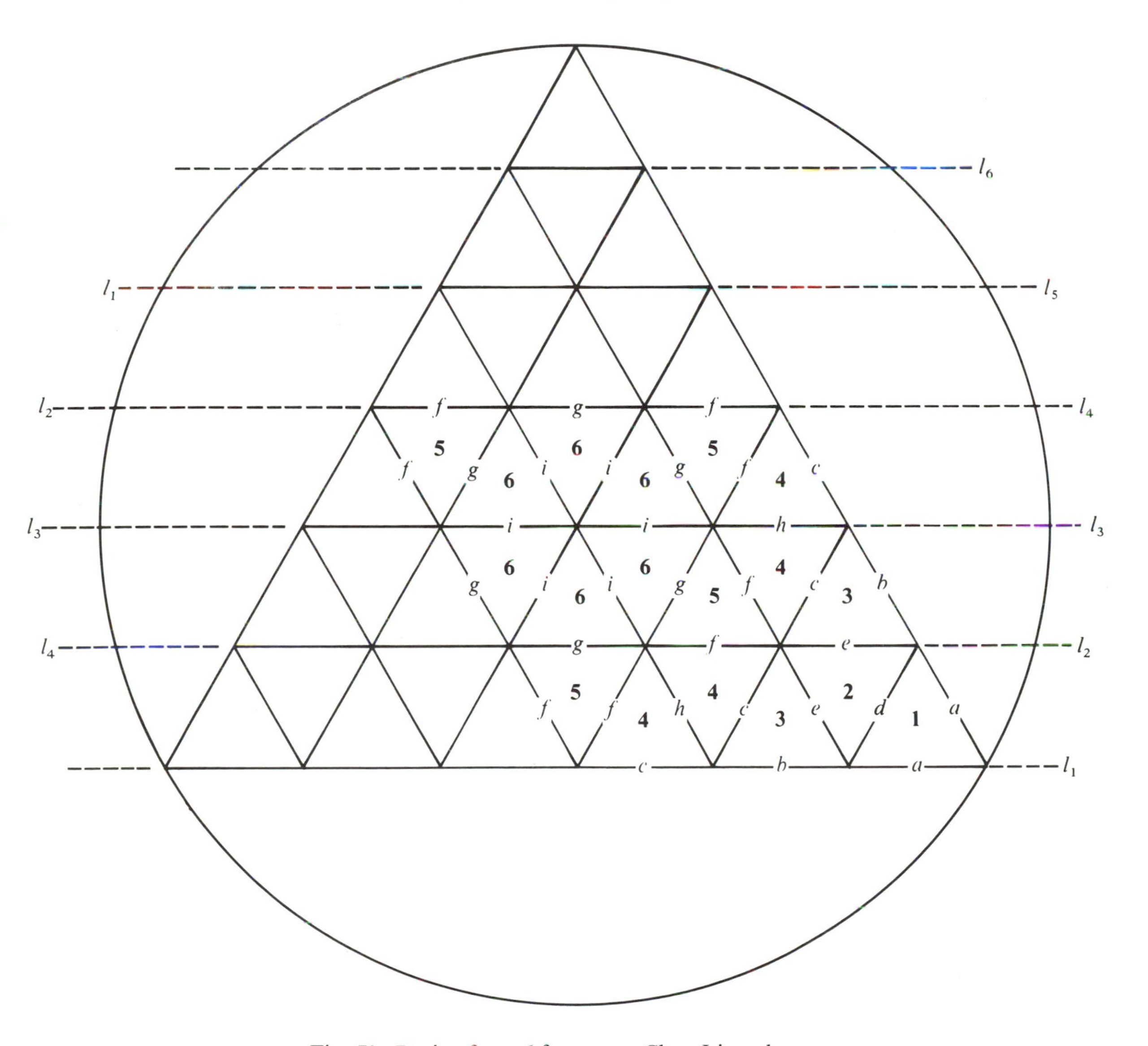

Fig. 70. Design for a 6-frequency Class I icosahedral lampshade. (Figs. 70a–70e. Gnomonic projections.)

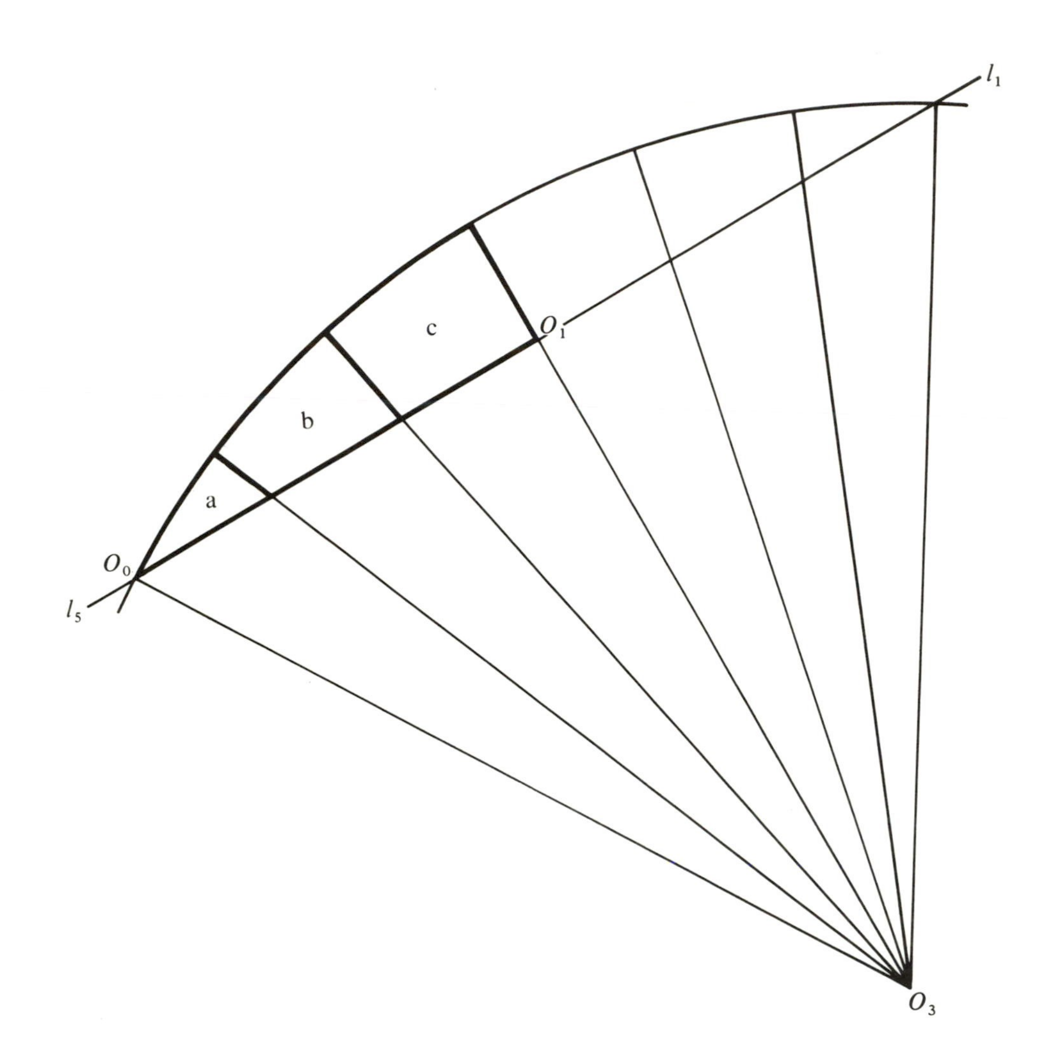

Fig. 70a.

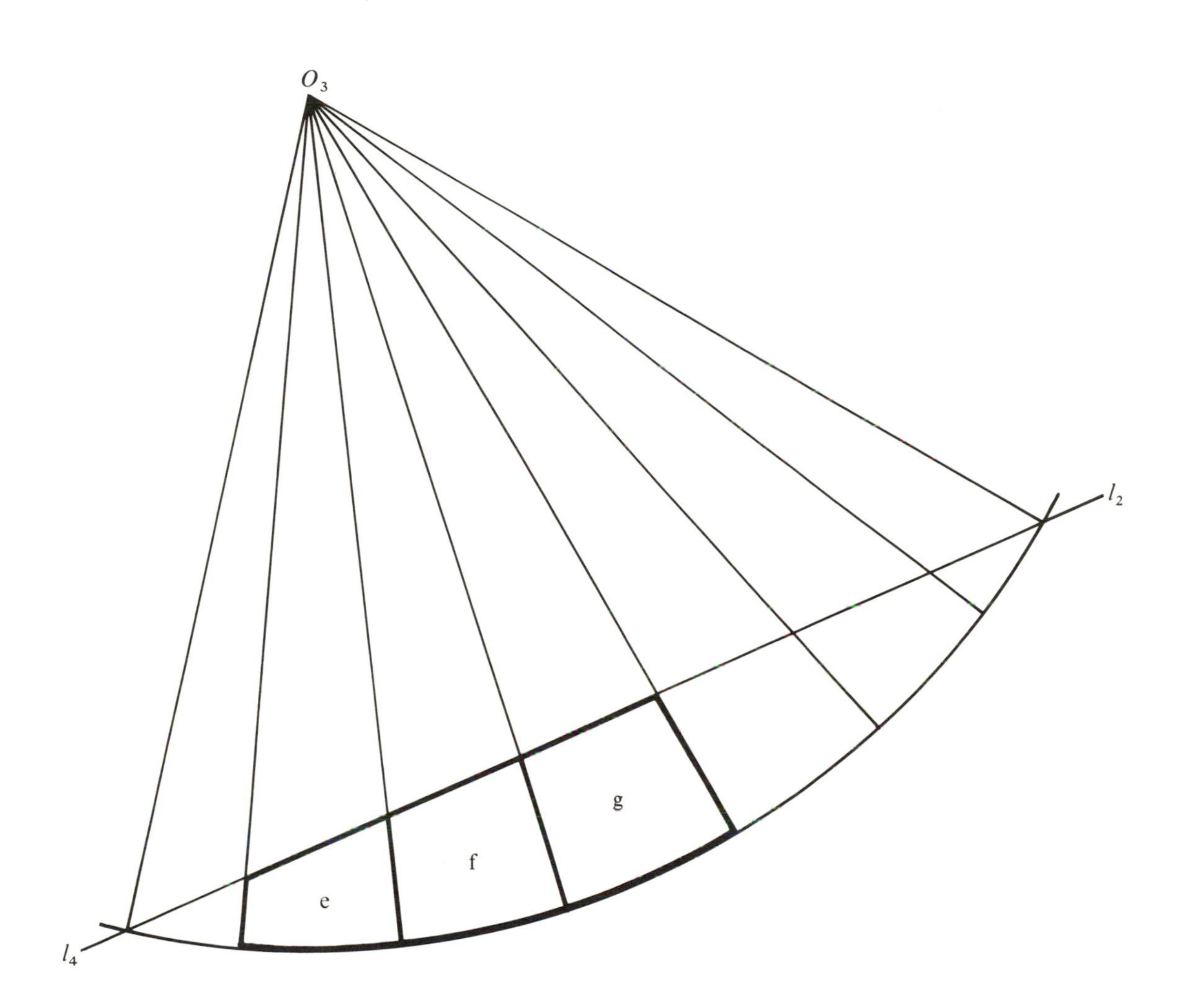

Fig. 70b.

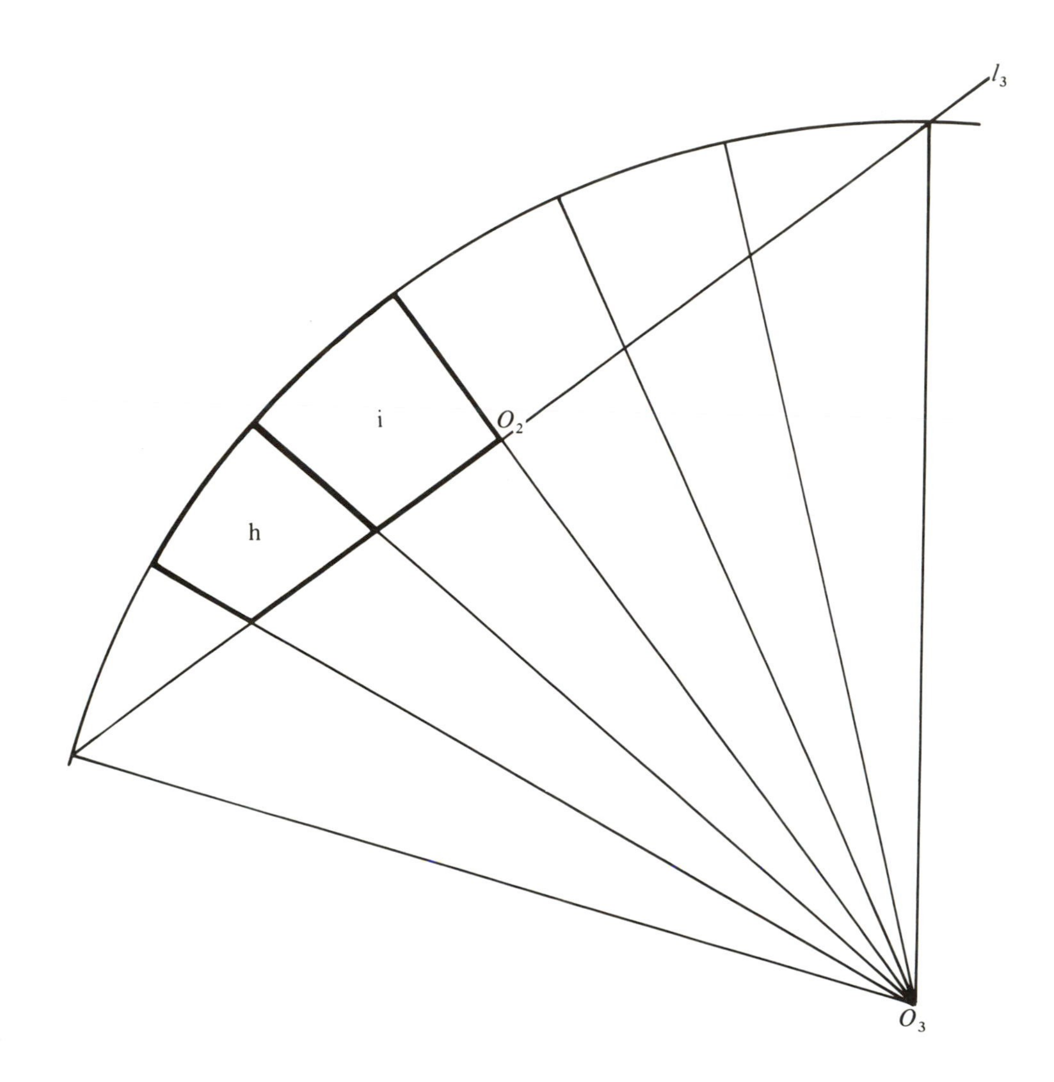

Fig. 70c.

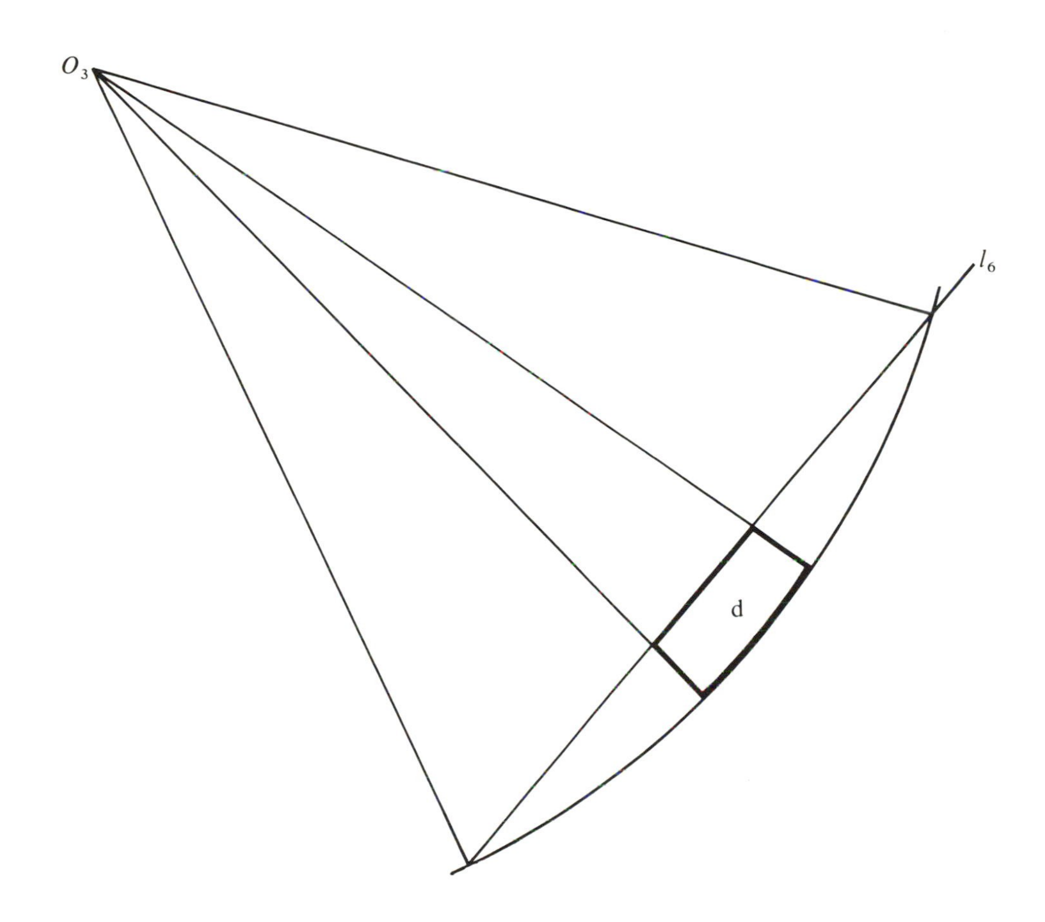

Fig. 70d.

Fig. 70e.

will easily change the arcs into corresponding chords. The chord lengths then become the lengths of the sides of the flat triangles needed to fill the spaces.

Chord lengths here are equivalent to what Clinton calls chord factors or strut lengths in *Domebook 2,* and the arc measures are the angles these struts subtend at the center of the dome. Clinton names this central angle δ and uses in its place for practical purposes in large-scale construction the so-called axial angle $\Omega = (180 - \delta)/2$, which is the angle between a strut and a radial line to the center. To change from central angles, as given in this book, to chord factors, as given by Clinton and also by Kenner and Pugh, you may use the formula: $cf = 2 \sin \frac{1}{2}\delta$, where cf stands for chord factor. A *chord factor* is defined as the measure of a chord in a sphere of unit radius. The δ of the formula may be replaced by χ, ϕ, ψ, the central angles used throughout this book.

Knowing these facts you can see how any spherical model or any geodesic dome given in Section III or Section IV can be given the flat-face treatment. Even the star-faced spherical models can be reduced to flat triangular elements. Think of the architectural richness to be achieved in this fashion!

Here is one last sample to bring this material on spherical models to a close. Figure 70 shows the icosahedron face in a 6-frequency segmentation. Figures 70a–70e suggest a way of designing a flat triangle bottom on a shallow cup whose upper lip remains curved. These cups take the place of the circular bands used on geodesic domes as previously made. This design has the effect of filling the spaces between the circular bands. The sphere need not be completely closed in. If a pentagonal section of suitable size is omitted near the end of the work, the inside remains open and visible, clearly revealing the flat icosahedral faces on the interior. If the paper is translucent this model can become a very attractive lighting fixture or lampshade, lighted from inside by an electric light bulb. If each of the six different kinds of cups is given its own color, with permutations of color from one icosahedron face to another, the resulting spherical mosaic of color is truly spectacular.

Epilogue

The spherical models in this book should provide you, first of all, with enough examples so that you may spend many hours making no more than what is illustrated here. But the study of this book should also stimulate you to construct other models, using designs of your own creation. Hopefully the book may also encourage you to study more deeply the mathematics involved in both the classification and the design of these forms.

The list of References shows that many books are available on this topic, most of them going far more deeply into the mathematics of these shapes and into related symmetry considerations than has been done in this book. This mathematics involves analytical geometry with spherical coordinates, vector analysis, and group theory.

It is my wish that you may have many happy hours in both the practical and theoretical aspects associated with these beautiful spherical models.

References

Beard, Col. R. S. *Patterns in space*. Creative Publications, 1973.

Clinton, Joseph D. *Advanced structural geometry studies*. NASA report NGR-14-008-002, 1970.

Clinton, Joseph D. "Geodesic math." In *Domebook 2*. Shelter Publications, 1974.

Coxeter, H. S. M. *Introduction to geometry*. Wiley, 1961.

Coxeter, H. S. M. "Virus macromolecules and geodesic domes." In *A spectrum of mathematics,* ed. J. C. Butcher. Auckland University Press and Oxford University Press, 1972.

Coxeter, H. S. M. *Regular polytopes*. Dover, 1973.

Coxeter, H. S. M. *Regular complex polytopes*. Cambridge University Press, 1974.

Coxeter, H. S. M., and Ball, W. W. R. *Mathematical recreations*. Dover, 1975.

Coxeter, H. S. M., Longuet-Higgins, M. S., and Miller, J. C. P. Uniform polyhedra. *Phil. Trans* 246A (1954): 401–50.

Critchlow, Keith. *Order in space*. Viking, 1969.

Cundy, H. M., and Rollett, P. P. *Mathematical models*. 2nd ed. Oxford University Press, 1961.

Donnay, J. D. H. *Spherical trigonometry*. Wiley (Interscience) 1945.

Heath, T. L. *Euclid's elements,* 3 volumes. Dover, 1956.

Hogben, Lancelot. *Mathematics for the million*. Norton, 1943.

Holden, Alan. *Shapes, space and symmetry*. Columbia University Press, 1971.

Kenner, Hugh. *Geodesic math and how to use it*. University of California Press, Berkeley, 1976.

Lines, L. *Solid geometry*. Dover, 1965.

Loeb, Arthur L. *Color and symmetry*. Wiley, 1971.

Loeb, Arthur L. *Space structures, their harmony and counterpoint*. Addison-Wesley, 1976.

Pugh, Anthony. *Polyhedra, a visual approach*. University of California Press, Berkeley, 1976.

Pugh, Anthony, *Introduction to tensegrity*. University of California Press, Berkeley, 1976.

Stewart, B. M. *Adventures among the toroids*. Published by the author, 1970.

Toth, L. Fejes. *Regular figures*. Pergamon, 1964.

Wachman, A., Burt, M., and Kleinmann, M. *Infinite polyhedra*. Technion, Haifa, Israel, 1974.

Wenninger, M. J. *Polyhedron models*. Cambridge University Press, 1971; paperback, 1974.

Wenninger, M. J. *Polyhedron models for the classroom*. 2nd ed. N.C.T.M. Publication, 1975.

List of models